EARTH MAGNETISM

Problems With Electric Charges, On Earth, In Atmosphere, In Van Allen Belt And On The Moon

NGUYEN VAN CUONG

Copyright © 2019

All rights reserved.

ISBN: 9781799165644

TEXT COPYRIGHT © [NGUYEN VAN CUONG]

All rights reserved. No part of this guide may be reproduced in any form without permission in writing from the publisher except in the case of brief quotations embodied in critical articles or reviews.

Legal & disclaimer

The information contained in this book and its contents is not designed to replace or take the place of any form of medical or professional advice; and is not meant to replace the need for independent medical, financial, legal or other professional advice or services, as may be required. The content and information in this book have been provided for educational and entertainment purposes only.

The content and information contained in this book have been compiled from sources deemed reliable, and it is accurate to the best of the author's knowledge, information, and belief. However, the author cannot guarantee its accuracy and validity and cannot be held liable for any errors and/or omissions. Further, changes are periodically made to this book as and when needed. Where appropriate and/or necessary, you must consult a professional (including but not limited to your doctor, attorney, financial advisor or such other professional advisor) before using any of the suggested remedies, techniques, or information in this book.

Upon using the contents and information contained in this book, you agree to hold harmless the author from and against any damages, costs, and expenses, including any legal fees potentially resulting from the application of any of the information provided by this book. This disclaimer applies to any loss, damages or injury caused by the use and application, whether directly or indirectly, of any advice or information presented, whether for breach of contract, tort, negligence, personal injury, criminal intent, or under any other cause of action.

You agree to accept all risks of using the information presented inside this book.

You agree that by continuing to read this book, where appropriate and/or necessary, you shall consult a professional (including but not limited to your doctor, attorney, or financial advisor or such other advisor as needed) before using any of the suggested remedies, techniques, or information in this book.

Table of Contents

INTRODUCTION 6

PART I: SOME VIABLE CONCEPTS AND THE SEA VS LAND CONTRIBUTIONS TO EARTH MAGNETISM 9

CHAPTER I: VIABLE RELEVANT CONCEPTS 9

I-Static Electric Charge and Coulomb: 9

II-A. Ampere: 12

III-Biot-Sawart laws: 15

IV-Major historic notes: 20

V-Major Features & compositions: 28

PREAMBLE 34

Author's Letter 53

CHAPTER II: AIR-EARTH CURRENT AND E.M 55

I-General About Atmosphere And Its Electrics: 55

II-Air-Earth Current In General (J_{cd}): 60

III-Scrutiny On Paradox, Questions And Causality: 66

CONCLUSIONS: 85

CHAPTER III: PROBLEMS OF SEA VS LAND CHARGES 87

I-PROBLEM OF A SINGLE CHARGE ON THE EARTH SURFACE: 88

II-PROBLEM OF OCEANS VS CONTINENTS 95

PART II: EXTERNALITY OR ASTRONOMY ON E.M 112

CHAPTER I: VAN ALLEN BELTS 114

CHAPTER II: TWO OPPOSITE HEMI-SPHERES OF MOON 123

CHAPTER III: MOON'S TOTAL CHARGE EXERTING ON EARTH 146

CHAPTER IV: TWIN CONES AT POLES 174

I-Earth's Poles: 174

II-Twin Cones At Poles: 175

III-Polar Fields At North And South: 187

IV-Light Intersection Near Poles And Pole Light: Aurora 191

CONCLUSIONS 195

Supplementary Images 208

CONCLUSION 219

INTRODUCTION

Dear Readers,

I study and compile my research results in a book. My purpose is to offer my book for publishing in Singapore and distributing to the world.

The book is sorted as Earth Basic Science; semi-empirical research, its title is "Earth Magnetism- some problems with atmosphere, Van Allen Belt and Moon".

Since 70th years in Marine University, I did battle with the issue of Earth Magnetism, but tens of questions are left behind when I left that school. Such long time of more than 30-year period seems not enough for me to consider the problems of Earth Magnetism, but I do come back with Earth Science and study Earth Magnetism again.

Should we begin with "the spinning & magnetism"?

This question has been asked for long ago but known to the world since 1915 when Samual J.Barnet (an American scientist) introduced his invent: the Barnet effect. The effect is stated in general: any neutral object (no electric charge on it) spins to induce magnetism.

That scientist, the author of his invent applied such theory to explain the origin of Earth Magnetism. He failed or is not accepted.

We now begin again with Earth rotation and its magnetism; nonetheless we don't introduce a shock in science, but we do introduce a new view to Earth Magnetism: it spins with electric charges on it cum in the electric field of some others, that's how it is magnetized.

Certainly we are at a large gap away from Samual J. Barnet. The gap is not just 102-year interim; it is a view to magnetism in general and Earth Magnetism in my research. The gap is presented in this book with several chapters in 200 pages.

I am neither going to deny the settled theories about E.M nor to make the Geomagnetism up-side-down, but consider the E.M as one constituted by 2 major parts:

- The "part I" for sea & land: This mainly comprises 2 compositions that induced by the charges at sea surface and land surface, they are both on Earth.

The very special paradox is that "land charges versus sea surface charge in contributing to the E.M whilst they both are moving together with Earth's rotation".

- The other-external inducers: is one that contributed by external magnetic inducers such as Van Allen belt, Moon and atmosphere. The simple principle governing this part is that the Earth is rotating in electric fields of many charges; that's how the Earth is magnetized. Meantime, in this part I point out that the Moon is charged and its charge induces magnetism to the Earth during orbiting around our planet in a cycle of 29.5 × 24 hrs.

Like any other authors, I wish my book to be printed and sold. Unlike some other author, my book is my claim about my discovery.

Thank you for your kind attention,

Ha Noi 09 Sept. 2017

Nguyen Van Cuong

PART I: SOME VIABLE CONCEPTS AND THE SEA VS LAND CONTRIBUTIONS TO EARTH MAGNETISM

CHAPTER I: VIABLE RELEVANT CONCEPTS

Nowadays, the quantum theory is the first and the most that catch the attention of the people in the world. Nonetheless, we should recognize that the physics can't be just bracketed in quantum, and so we got to look backward some centuries when we can find W. Coulomb, A.Ampere and the two others J.B.Biot and F.Savart.

I-Static Electric Charge and Coulomb:
This is the first we inspire when discuss about Earth Magnetism, the subject is attached to a scientist who discovered electric charge: W. Coulomb.

His rule is simply expressed in a formula for the force (F) between the two charges:

$F = k_e * q_1 * q_2 / r^2$

Or

$= k_e * q_1 * q_2 * r^{-2}$

Where q1 and q2 are values of charges, each can be either (-) negative or (+) positive. The distance between the two is "**r**" and the Coulomb constant "**k_e**" is depending on the media between the two.

Ke = 8.99*10⁹ N.m².C⁻²: Coulomb constant for the normal environment or vacuum between the two charges. The force is found to be negative (-) when either q1 or q2 is negative; to be positive (+) when both of them are positive or negative; this is quite a law in math and it demonstrates the wise of force "+ push" or "- pull".

Portrait of Charles-Augustin de Coulomb

Charles-Augustin de Coulomb was an eminent French physicist. He formulated the Coulomb's law, which deals with electrostatic interaction between electrically charged particles. The coulomb, SI unit of electric charge, was named after him.

We are considering "ke" for the media of large scale and complicated, therefore we should give out some assumptions before any problem.

Some more about moon-earth relation and calculation are debated in the "Moon charge and E.M" of this book.

II-A. Ampere:

The second scientist we inspire is A. Ampere, he is a great scientist but simply his rule should be taken as: total charge in Coulomb transported through a point within a unit of time (second) is to make current value one Ampere.

Portrait of André-Marie Ampère

André-Marie Ampère (1775-1836) is a French Physicist and Mathematician; He worked at the beginning of the 1800s in Paris, France. He used his mathematical and statistical skills to observe and measure natural occurrences that had been discovered by other European scientists. He went on to produce complete proof of the relationship between electricity and magnetism. He also developed a new way of classifying chemical elements. Amp (Ampere) or his name is given to basic unit of electricity.

With the above concept of electric current, an instant of the 5 different charges moving through a point in 5 seconds can make an average current in that interim as following:

2 Coulombs in 1st second;

3 Coulombs in 2nd second;

1 Coulomb in 3rd second;

2 Coulombs in 4th second;

4 Coulombs in 5th second, can make a current of:

$$I = \frac{2+3+1+2+4}{5} = 2.4 \; Amperes$$

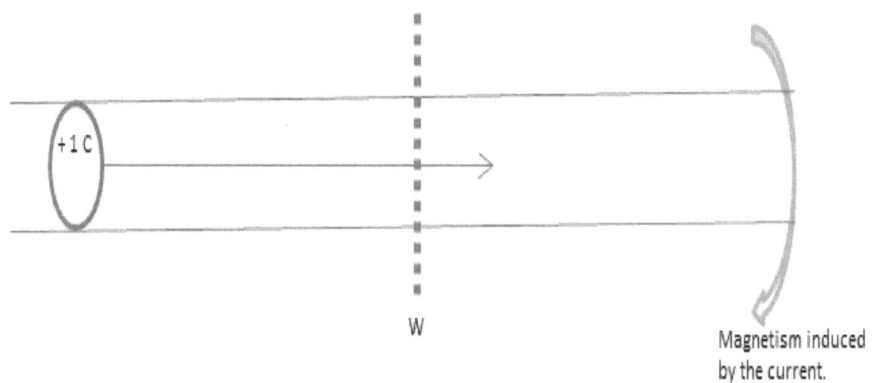

Figure 1/I-Electric current & magnetism around

As soon as an electric current is created, a magnetic field is established around it.

III-Biot-Sawart laws:

In physics, the momentum as well as the "particle spin" is considered as the most interesting subject. Let's leave the modern physics and work with these two scientists.

Jean Baptiste Biot – a French scientist (1774-1862)

Felix Savart-a French physicist (1791 -1841)

Those two French physicians are known as inventers to the law that named after the both, the Biot-Savart about electromagnetism.

1-<u>Biot - Savart law</u>: The kernel of the law is to say that a charge motion is to make magnetic field. A point "P", at a distance "r" from a conducting wire where the charge moves in, is in a magnetic field which can be calculated as following:

$$\boxed{d\vec{B} = \frac{\mu_0}{4\pi} \frac{I\, d\vec{s} \times \hat{r}}{r^2}}$$

$$\vec{B} = \frac{\mu_0}{4\pi} \frac{q\vec{v} \times \hat{r}}{r^2}$$

Where dB is additional magnetic field contribution induced

by the current **"I"**, which is created by the directional move of a charge **"q"**.

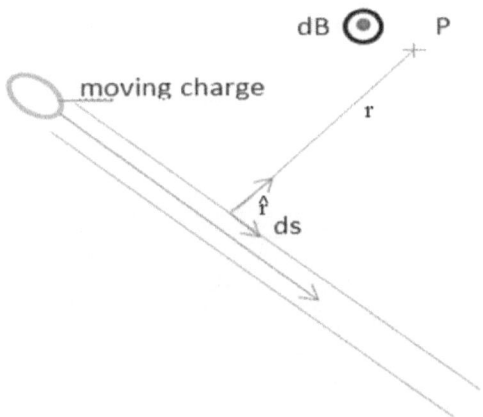

Figure 2/I-A charge move and its magnetism contribution to point P.

And **"ds"** is a small segment of wire, both **"ds"** and **"r̂"** are respective vector units of wire length and distance from the wire to **"P"**; and product of those is dB -an outward vector or the vector that comes toward reader as following figure.

μ_0 – permeability coefficient of environment, is $4\pi * 10^{-7}$ in normal air or vacuum.

2-Biot-Savart problem of many moving charged points:

Our discussion is not the first but is the most interesting

application of Biot-Savart law, which will be applied to almost every issue in this book. Instead of momentum and spinning, we work with charge and orbiting. The following is an excerpt from a manual on web.mit.edu.com (Mit. University web).

Quote:

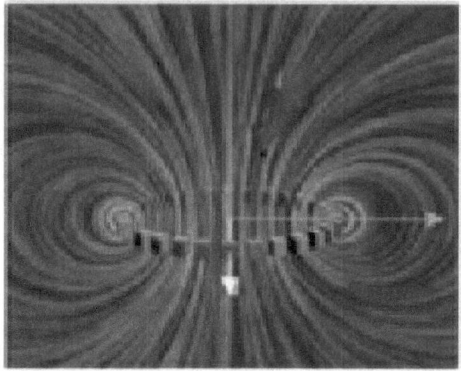

The magnetic field due to 30 charges moving in a circle at a given observation point. The position of the observation point can be varied to see how the magnetic field of individual charges adds up to give the total field.

In the above figure (9.1.11), we show an interactive Shock Wave display that is similar to that in figure 9.1.10, but now we can interact with the display to move the position of observer about in space. To get a feel for total magnetic field, we also show a "iron filings" representation of the magnetic field due to these charges. We can move the observation point about in space to see how the total field at various points arises from the individual contributions of the magnetic field of to each moving charge.

$$\vec{B} = \frac{\mu_0}{4\pi} \frac{q\vec{v}\times\hat{r}}{r^2} \qquad (9.1.20)$$

Note, however, that since a point charge does not constitute a steady current, the above equation strictly speaking only holds in the non-relativistic limit where v<<C, the speed of light, so that the effect of **"retardation"** can be ignored.

The result may be readily extended to a collection of N point charges, each moving with a different velocity. Let the ith charge qi be located at (x,y,z) and moving with velocity v, using the superposition principle, the magnetic field at P can be obtained as:

$$\vec{B} = \sum_{i=1}^{N} \frac{\mu_0}{4\pi} q_i \vec{v}_i \times \left[\frac{(x-x_i)\hat{i}+(y-y_i)\hat{j}+(z-z_i)\hat{k}}{\left[(x-x_i)^2+(y-y_i)^2+(z-z_i)^2\right]^{3/2}} \right] \qquad (9.1.21)$$

Unquote.

IV-Major historic notes:

This book is not written about industrial magnetism but the subject is vital to the contemporary life, therefore the following note must benefit the readers in general. On the other hand, some reader can't get through this book without reference back these notes.

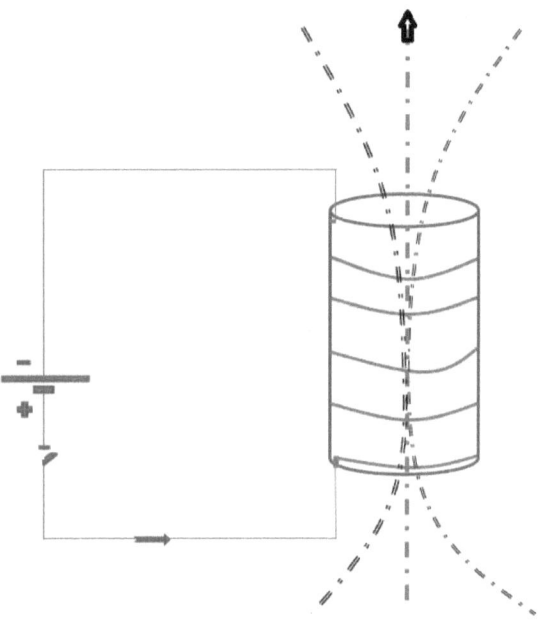

The above figure illustrates how a direct current induces the magnetism during flows through a coil.

1-Magnetic field intensity: This is defined as the ratio between the max magnetic flux density of a circuit to the permeability of free space.

H=B₀/μ₀

The magnetic intensity depends on the geometry of the circuit and the conduction current, not on the medium.

The total magnetic field in a magnetic material is due to the conduction current in the external circuit and the microscopic current developed in the material.

B = B₀ + μ₀*M

2-Field strength (H):

Definition: It is the force experienced by a unit north pole at a point in a magnetic field

H = F/m

Where F is the force, m is pole strength.

- If the field lines are more, then H is high and vice versa.

- Thus, the closer to the pole, the higher strength of field.

Nowadays, magnet is applied in almost every equipment or even toy. The power of U-shaped magnet is known to everyone who is experienced with machinery work.

3-Magnetization:

a-Magnetic domains in an un-magnetised bar point to random directions.

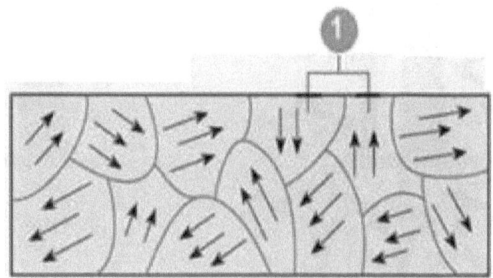

b-A bar magnet is brought near the un-magnetised bar.

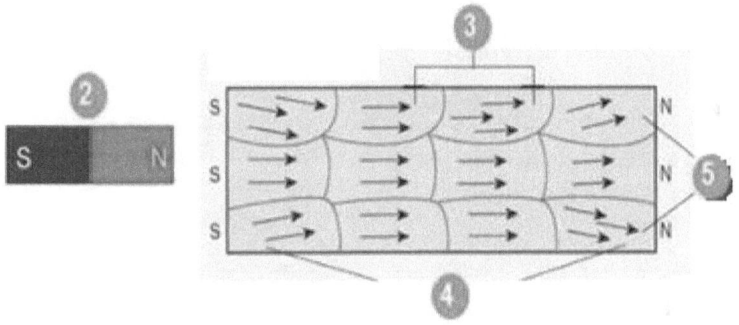

c- The magnetic domains point in the same direction, producing magnetism. N or S poles of adjacent domains cancel each other out.

d- The atomic magnets at both ends are free. This produces N and S poles at the ends.

e- The atomic magnets at the ends tend to fan out due to repulsion between the like poles.

The magnet and its force applied in industry are not magic; they are available in every material and everything, their force is null because every elementary magnet in there points to different directions or even one eliminates the other. The external magnetic force helps a certain bar in its adjacent area to re-direct every elementary direction and make a force of their total.

4-Magnetic Equivalent Circuit:

<u>Magnetic Field Intensity (H):</u> the amount of field force (mmf) distributed over the length of the electromagnet. Sometimes referred to as Magnetizing Force.

$$B = \mu H$$

$$H = \frac{Ni}{l} \, (At/m)$$

$$B = \frac{\mu Ni}{l} \, (T)$$

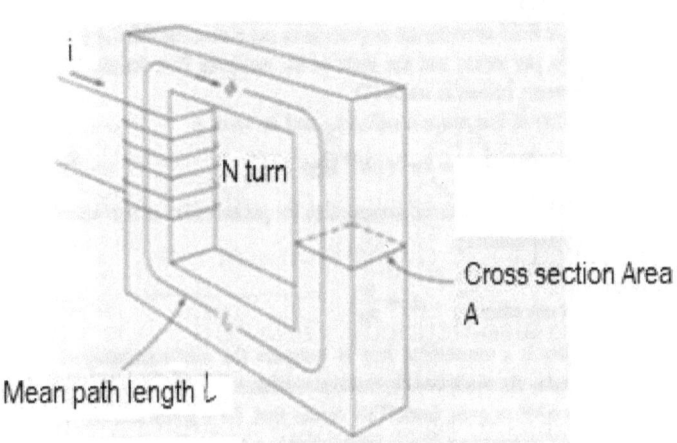

<u>Magnetic Flux Density (B)</u>-The amount of magnetic field flux concentrated in a given area.

Where N-numbers of turns of coil; i-current in the coil; H-magnetic field intensity; l-mean length of the core.

<u>Relation between B and H:</u>

The magnetic field intensity, H produces a magnetic flux density, B everywhere it exists.

B = µH (weber/m²) or Tesla

B = (µ$_r$/µ$_0$)H (w/m²) or T

µ: Permeability of the medium.

µ$_0$: Permeability of free space=$4\pi.10^{-7}$ wbA.t.m

$\mu_r = \mu/\mu_0$ relative permeability of medium.

Magnetic Flux Φ

-Unit for flux is "weber"

-The definition of 1 weber is the amount of flux that can produce and induced voltage of 1 volt on a one-turn coil if the flux reduced to zero with uniform rate.

This concept is also defined as following: It is a measure of magnetic field in a certain medium. In simple term if the magnetic field has to pass through a certain medium, it will always travel as "flux" lines (flux lines are imaginary, but continuous lines travelling from North pole to South pole of a magnet. Its units are Weber.

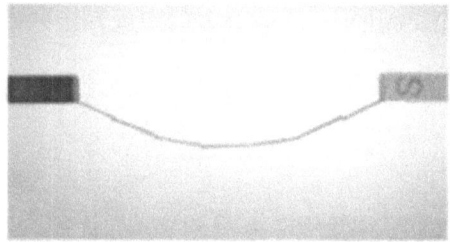
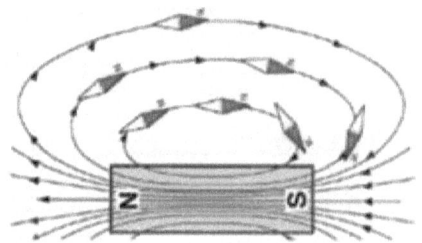

Magnetic Flux Density B

-Unit for magnetic flux density is Tesla.

-The definition of 1 tesla is the flux density that can

produce a force of 1 Newton per meter acting a conductor carrying 1 ampere of current.

Magnetic Field Strength H

-Unit for magnetic field strength is Ampere/meter.

-A line of force that produce the flux.

Earth's Magnetic intensity in history and now (an excerpt):

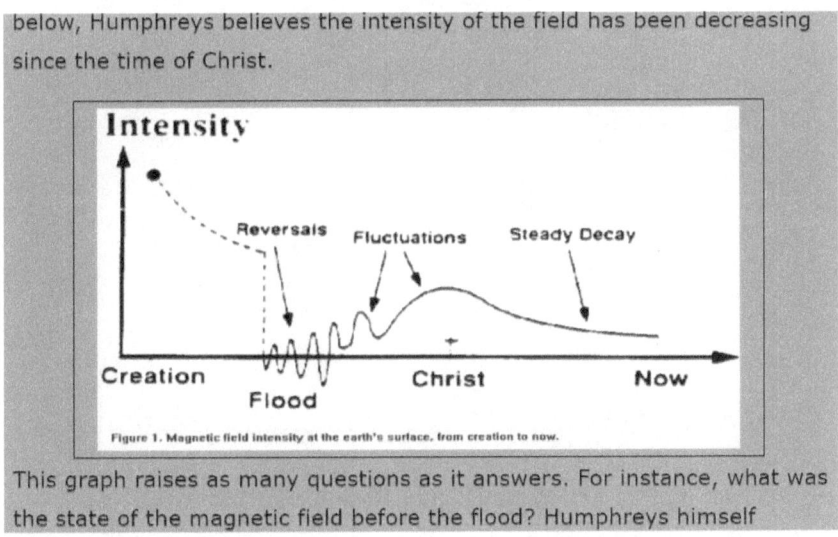

below, Humphreys believes the intensity of the field has been decreasing since the time of Christ.

Figure 1. Magnetic field intensity at the earth's surface, from creation to now.

This graph raises as many questions as it answers. For instance, what was the state of the magnetic field before the flood? Humphreys himself

The above note is cited as some information about the Earth's magnetism in history and now. Definitely Humphreys does not fabricate the graph, and he draws it with facts that found in researching. The question is "steady decay" now and then?

V-Major Features & compositions:

1-E.M polarity: at present, south magnetic pole is nearly coinciding with earth north geographic pole, that's why magnetic compass north hand always points toward North.

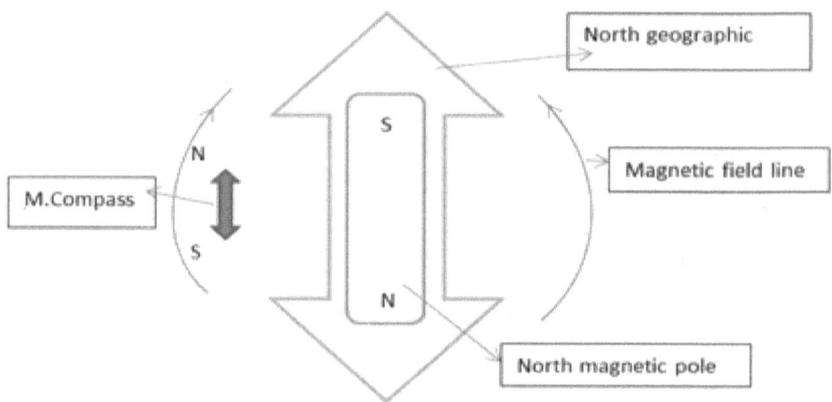

Figure-3/I-Earth Magnet with zero-deviation.

The above image is set with assumption that magnetic deviation is "nil", and South magnetic pole is on North geographic pole.

2-Earth Magnetic field in different coordinates:

The following is an excerpt from manual of Bremen University

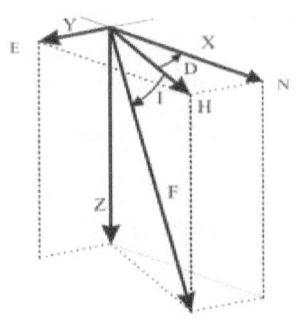

Local cartesian coordinates (X, Y, Z)
- X: northward component,
- Y: eastward component,
- Z: downward component.

Local cylindrical coordinates (H, D, Z)
- $H = \sqrt{X^2 + Y^2}$: horizontal intensity,
- D: declination (deviation from true north = geographic north).

Local spherical coordinates (F, I, D)
- $F = \sqrt{X^2 + Y^2 + Z^2}$: total intensity,
- I: inclination (magnetic dip angle).

[(1) Glaßmeier et al.]

3-Earth Magnetic field compositions:

Earth Magnetism or E.M is contributed by several magnetic inducers, we are going to discuss about someone and not the whole. Unlike traditional way, our way is to start from inducers such as charge on continent, sea surface charge and Moon charge. Therefore we got to denote each magnetic composition with a number attached to a symbol "Δ" such as Δ(1) or Δ(2), except B(e) which is induced by inner rotor of the Earth as well as local man-made source. B(e) is also denoted as Δ(7), the only one composition which has undefined direction and left as non-directional quantity, while all the rest are directional.

- Positive charge on land contribution: Δ(1), note that positive as well as negative charge is everywhere, but we do not consider every one as it is but with massive quantity as representative of each. Therefore a continent is positively charged in general. Direction of Δ(1) will be defined in its problem, the land charge contribution in chapter II of part I-"sea surface charge versus land charges".

- Negative sea surface contribution: $\Delta(2)$, unlike continental surface which is positive, the sea surface is homogenous as only one electrode and always negative. The direction of this composition is defined in its problem, chapter II of part I.

- Van Allen belt contribution: $\Delta(3)$, Van Allen belt and its companion, the plasma sphere, are not just making a shield to protect the Earth, they are charged and to contribute to E.M. The direction of their contribution is defined in the problem in chapter I of part II.

- Negative moon's hemisphere contribution: $\Delta(4)$, the Moon primarily is assumed as neutral with 2 hemispheres of 2 opposite charges. Each hemisphere is inducing magnetism to Earth different to the other. The direction of this quantity is defined in its problem in part II.

- Positive Moon's hemisphere contribution: $\Delta(5)$, the shined hemisphere of Moon is positive as losing electrons under solar wind. It induces magnetism to Earth and its direction is defined in its problem in part II.

- Contribution of total charge of Moon: $\Delta(6)$, the above mentioned problems are considering Moon as neutral of 2 opposite charges, but in reality the total of those charges $Q(a)+Q(b)$ rarely is neutral, besides a certain value different to Zero; therefore its contribution is considered and it has potential to be a huge surprise.

V-Two important characteristics of material:

These are not just 2 important characteristic of material, they are 2 subjects that attracts a lot of researchers in history and now. Although we can go along with the researchers but we can't keep going ahead without some brief about them.

1-Permeability: The magnetic permeability of a material is the degree of magnetization that it receives when it responds linearly to a magnetic field. The symbol denoted for permeability is a Greek letter "µ", which is measured in units of H/m (Henry per meter) or N/A^2 (Newton per Ampere squared). Empty space itself has an electromagnetic permeability value, also known as the magnetic constant, $\mu_0 = 4\pi \times 10^{-7} \, N/A^2$

For the purpose of demonstrating a mathematical relationship between µ and some magnetic field B that is influencing a medium of some king, let's assume that there is an auxiliary magnetic field H that represents the way this B field influences the organization of magnetic dipoles in our medium. According to the way we have set up situation, the relationship between the fields B, H and our permeability µ is mathematically defined by the following equation:

B = µH

In the above case, our permeability will be scalar as long as our medium is isotropic. Isotropy refers to a certain organization of medium that means it is uniform in all directions. A good example of the use of isotropy is in the cosmological principle that is a part of the Big Bang Theory. It is assumed, in the Big Bang model, that space is isotropic, meaning there are no localized anomalies in space that would stop the universe from expanding in the way we observe. Anisotropy is opposite of isotropy. We are able to use a scalar for permeability if we are working with a second rank tensor for anisotropy linear medium.

2-Susceptibility: Magnetic susceptibility is the degree of magnetism of a material in response to an applied magnetic field. If the magnetic susceptibility is positive, the material can be paramagnetic, ferromagnetic or antiferromagnetic. In these cases the magnetic field is strengthened by the presence of the material. Alternatively, if the magnetic susceptibility is negative, the material is diamagnetic and as a result, the magnetic field is weakened by the presence of the material.

If the ratio between the induced magnetization and the inducing field is expressed in unit volume, volume susceptibility (k) is defined as

K = M/H

Where M is the volume magnetization induced in a material of susceptibility k by the applied external field H. Volume susceptibility is a dimensionless quantity. The value depends on the measurement system used:

K(SI) = 4π*k(cgs) = 4π*G*Oe^{-1},

Where G and Oe are abrreviations for Gause and Orstedt respectively. The SI system should be used.

Mass, or specific, susceptibility is defined as

X = k/r

Where r is the density of material. The dimensions of mass susceptibility are therefore cubic meters per kilogram.

PREAMBLE

Earth Magnetism or E.M is one of the oldest themes in the Earth study. The world scientists are researching, debating about the same theme for several decades; I assume that there is not only one but some of them even dropped it.

I am neither going to deny the settled theories about E.M nor to make the Geomagnetism up-side-down, but consider the E.M as one constituted by 2 major parts. The "part I" mainly comprises underground rotors near the Earth core, the charges at sea surface and land surface. The other is one that contributed by external magnetic inducers such as Van Allen belt, Moon and atmosphere.

The mechanism in magnetic inducement of E.M is rather complicated and likely to be ruled by some principles that had not been mentioned so far; each spot on the Earth is different to any other.

The very special paradox is that "land charges versus sea surface charge in contributing to the E.M whilst they both are moving together with Earth's rotation". Although land charges and sea surface charge are influenced by external movement, I do keep them in a separate part (Part I) for a better discussion.

Actually, I leave what happen inside the Earth as the stuff of a conservatory; just add something on the settled basis with some effects on Earth surface and from sky.

Many scientists assume that the astronomic objects may contribute several per cents in the E.M. Here in this book, I argue that the contribution from those might go beyond any previous prediction and certainly it must not be such tiny as assumption of those scientists.

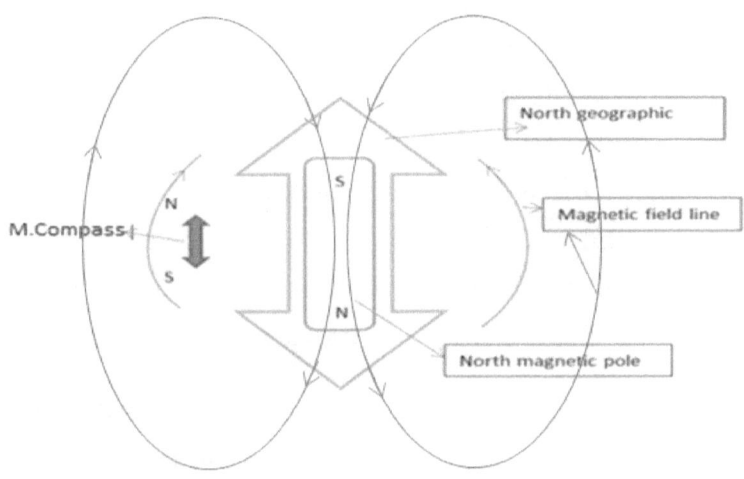

Figure 1/F-Earth Magnetism is opposite to geographic polarization

E.M is a giant theme like the huge size of the Earth. I am capable and proud to interpret some of the information that compiled by many scientists in history and especially in about a century time.

Furthermore, like any other semi-empirical study, this book presents some latest E.M data and I analyse them with my own view. On some aspect, this book is more likely presenting a discussion by which I contribute my own view to a theme that researched by tens of scientists in the world.

Like any other book, this is compiled with reference to many books as well as researching reports from many scientists in the world. The knowledge from electricity to atmosphere, from math to physics; all of them are referred to facilitate the research.

The Earth rotation is definitely to suggest us that E.M is created by the rotation, or the Earth is viewed as an electromagnet. I trust that the E.M must look alike a normal magnet if we eliminate all external compositions. This assumption has been the subject of much and long study, particularly by S.J. Barnett, who has shown that a piece of iron may be magnetized by rotation.

In 1900, Sutherland suggested a possible cause of the Earth Magnetism the rotation of an electrostatic field within the Earth —a positively charged core and a negatively charged crust, or vice versa.

The later analysis made by Bauer in 1922, using improved data based on modern observations, gave approximately the same result.

Sutherlands, Bauer and Swann suggested that we may have to look for some slight but fundamental modification to those accepted laws.

(Magnetic storm) it was recognized that the more severe

magnetic disturbances occur at practically the same time all over the Earth, and further comparative study of abrupt beginnings and sharp turning points indicated strict simultaneity.

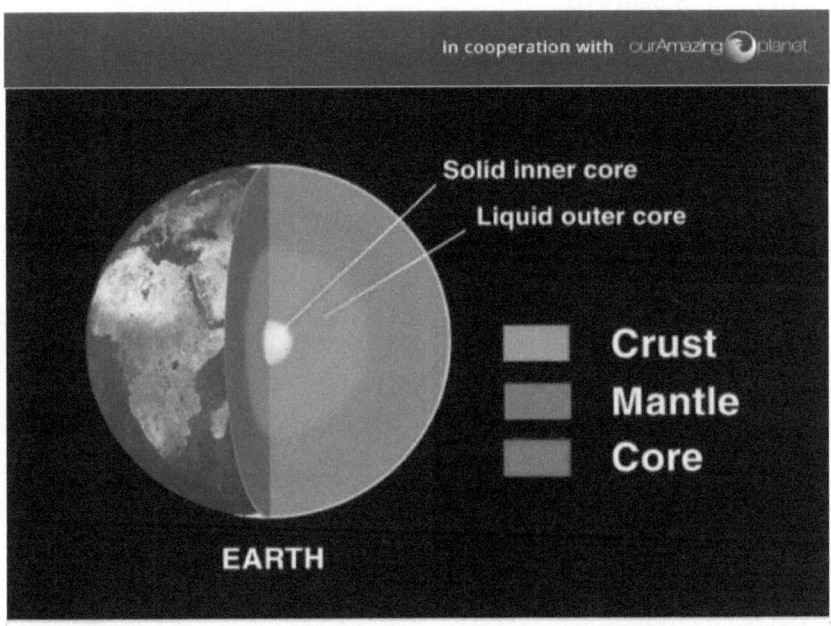

Earth has multiple layers: the crust, the mantle, the liquid outer core and the solid inner core.
Credit: NASA

Picture 2/F- Structure of the Earth

The world scientists keep researching the Earth as well as E.M at all time. The followings are some information in brief about the science achievements:

- Temperature at Earth's inner core boundary: by measuring the melting point of iron at high pressure (about 2.2 million times more than normal pressure on Earth surface) in laboratory, the scientists drew a result to calculate the temperature now estimated at 6,000 °C (about 10,800 F), that's as hot as the surface of the Sun?

- Earth's inner core is solid, while its outer core is molten or loose then crust and mantle above the outer core. Along with Earth's spin they create the magnetic field?

X

X   X

Before any further discussion about E.M, we should not neglect the following from an important book TERRESTRIAL MAGNETISM.

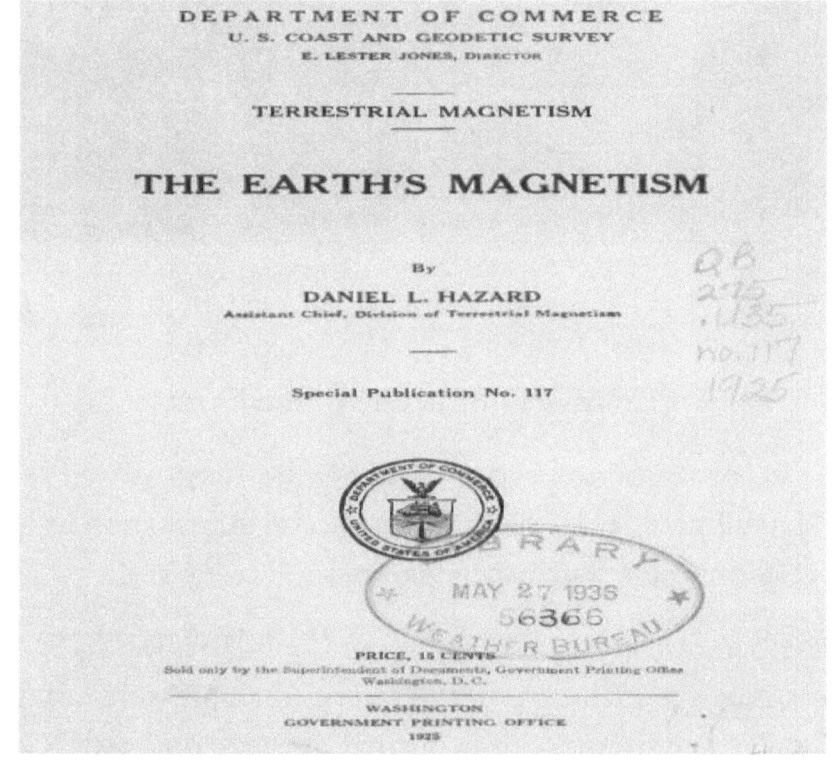

Image 3/F-Terrestrial Magnetism-1925

Quote:

In the previous centuries, the XIX and XX, human being had carried out a series of attacks to the Earth Magnetism, the questions are: What it is, What caused it, What cause it to change. Many leading physiscists, mathemeticians joined in those attacks. Many theories, one after another had been advanced then withdrawn if it was found inconsistent with some of the observed facts.

1/They are inadequate or inconsistent because:

- Can fit well with quality but can't do with quantity.

- Can fit well with primary stage but can't do with the next when our knowledge is improved.

2/The advances in other field of science have been employed:

- Cathod rays, electronic theory of matter.

- Constitution of the Sun.

- Earth interior.

3/Gilbert's conception: Earth is a great magnet uniformly magnetized about the Earth own rotating axis. This theory is discarded afterward when the high temperature inside the Earth is recognized by then some material is demagnetized in such high temparature condition.

4/Under high pressure: properties of some matter in great pressure may be materially different from those observed in normal condition.

Susceptibility to magnetization is one of those properties. Some geophysicists suggest that the Earth's inner core is composed by almost entirely Nikel and Iron and a possibility of susciptibility of magnetization in spite of high temparature, because of enormouse high pressure. Some magneticians are inclined to work further to consider The Earth as a giant magnet.

5/ Electric currents flowing around The Earth: When it seemed clear that the conception of the Earth as a permanent magnet could be sustained, the idea was advanced that the Earth magnetic field might be due to electric currents flowing about the Earth, either below the surface or in the atmosphere-the Earth an electromagnet.

6/ Three portions constituting the E-magnetism: The mathematical analysis of the Earth magnetic field, according to the method devised by Gauss and extended by Neumayer and Peterson (1891) and Schmidt (1896), indicated that:

- A portion of Earth Magnetism, <u>perhaps one-fortieth</u>, could be referred to forces outside the Earth.

- Another small portion to <u>vertical electric currents.</u>

- But by far the larger part to a system of <u>force within the Earth.</u>

<u>A new analysis made by Bauer in 1922, using improved data based on modern observations, gave approximately the same result.</u> He reached the conclusion that for a satisfaction representation of the observed data it is necessary to recognize the existence of an internal magnetic system constituting about <u>94 per cent of total field</u>, with an external system and a non-potential system each amounting to about <u>3 per c</u>ents.

A comparison of his results with those previously obtained for the epochs: 1842 and 1885 indicated that the intensity of magnetization of the Earth had been decreasing during the 80 years at an average annual rate of <u>one part in 1,500</u>; a rate of loss which it is hard to reconcile with the age of Earth and the present intensity of magnetization unless we suppose that there have been also periods of increasing intensity.

7/ E-Magnetism caused by Earth's rotation: The idea of Earth as an electro magnet naturally suggested the possibility that its magnetism may be caused by its rotation. This possibility has been the subject of much study, particularly by S.J. Barnett, who has shown that a piece of iron may be magnetized by rotation, though the <u>observed effect was much too small</u> to account for Earth Magnetism. In 1900, Sutherland suggested as a possible cause of the Earth Magnetism the rotation of an electrostatic field within the Earth —a positively charged core and a negatively charged crust, or vice versa.

8/ Electronic theory and Sutherland: The development of electronic theory of matter, with the atom consisting of a positively charged nucleus surrounded by negatively charged electrons, led Sutherland to suggest that if for some unknown reason, connected perhaps with gravitation, the negative charge of the atom is farther from centre of Earth than the positive charge by only 4×10^{-9} cm, it would account for a magnetic field comparable with that of the Earth. When the electronic theory had been more fully developed and Sutherland's hypothesis submitted to further test, it was found to be untenable either qualitatively or quantitatively.

Failing to find a satisfactory explanation to the Earth's magnetism on basis of the known properties of matters and accepted laws of *electrodynamics, Sutherlands, Bauer and Swann suggested that we may have to look for <u>some slight but fundamental modification of those accepted laws</u>, possibly as regard the mutual attraction and repulsion of moving positive and negative electrons, similar to a suggestion of Lorenz regarding the cause of gravitation. Indeed there seems to be growing a belief that gravitation and terrestrial magnetism are very closely allied and probably to be traced to a common origin.*

9/ <u>Atmosphere electric & Earth currents</u>: In view of difficulties in the way of a direct solution of the problem of the cause of Earth Magnetism, the magnetism researchers have turned their attention to a study of its variations and their correlation with associated phenomena, such as atmospheric electricity, Earth currents, aurora, sun spots, solar radiation, in the hope that the cause of variations may be discovered, and that in that way light would be thrown on the main problem. In particular, magnetic storms, those irregular disturbances of large amplitude and comparatively short duration, have been the subject of much study.

It frequently happens that the occurrence of magnetic storms and auroras coincides with the presence of large spots on the sun, and this naturally has led to attempts to trace a causal relationship. <u>It was soon seen that a direct magnetic effect by the sun was out of the question, because of great distance</u>. With the development of the idea that the Earth's magnetism may be caused by current of electricity, different forms of electric discharge emanating from the sun were successively put forth as the cause of observed terrestrial phenomena, the theories advanced keep pace with the development of our knowledge of electrical discharges in vacuum.

The correlation between magnetic storms with sunspots, although very *satisfactory when based on yearly average, leaves much to be desired when individual cases are considered. Thus severe <u>magnetic storms sometimes occur when no large sunspots are visible</u>, and on the other hand a sunspot is not always accompanied by a magnetic storm. Large magnetic storms always follow each other at an interval approximating the time of revolution of the sun and such reoccurrence has been traced with several rotation periods, <u>not every recurrence is accompanied by a sunspot, however</u>.*

10/ Maunder hypothesis:

To *meet these conditions, Maunder advanced the hypothesis:*

- The solar activity which gives rise to magnetic disturbances on the Earth does not act equally in all directions but along narrow well-defined streams, not necessarily truly radial.

- These streams arise from active areas of limit extent.

- These active areas are not only the source of our magnetic disturbances but are also the seats of formation of sun spots.

- These areas can be active both before a spot has formed and after it has disappeared.

Birkeland, Arrhenius, and Normannd agree to consider aurora as a luminescence produced by the absorption of cathode rays in the upper atmosphere and attracted toward the Earth magnetic poles. Birkeland, who devoted many years to the study of auroras, advances a theory that cathode rays from sun can set up electric currents in the atmosphere which in turn give rise to secondary cathode rays. He supported his theory by the production of artificial auroras in laboratory, about a magnetized steel ball in a tube of rarefied air exposed to cathode rays.

11/ Sun's electrified particle streams:

If magnetic disturbances on the Earth are to be ascribed to streams of electrified particles shot out from the sun, we should expect to find a disturbance occurring first at the part of the Earth first entering the stream and later at other places as they in turn entered the stream, as a result of the combined motion of the Earth in its orbit and the sun about its axis. From the time of the earliest comparison of photographic records from widely separate observatories, <u>it was recognized that the more severe magnetic disturbances occur at practically the same time all over the Earth</u>, and further comparative study of abrupt beginnings and sharp turning points indicated strict simultaneity, the departures therefrom being ascribed to errors inherent in the time measurements, so that <u>the more accurate determination</u> of the time of occurrence of such salient features was suggested as a method of determining differences of longitude."

Unquote

The latest document for the same theme-the Earth's magnetic field, is one from a French Scientist; I would like to cite here a short excerpt from his argument:

Quote

THE ORIGIN OF THE EARTH'S MAGNETIC FIELD: FUNDAMENTAL OF ENVIRONMENTAL RESEARCH? (Emmanuel Dormy-MHD in Astro-and Geophysics, Department De Physique, LRA, Ecole Normale Superieure-France.)

The origin of The Earth's magnetic Field is a long-standing issue, which has captured attention of many reknown scientists. If William Gilbert, Andre-Marie Ampere, Rene Descartes, Edmond Harlley, Kart Friedrich Gauss, Lord Brackett and many others who contributed to the development of science, have worked on this problem, it is mainly because it related to a very important issue of very critical importance: navigation at sea. This is not so true any more, now that satellites provide the precise latitude and longitude without the need for us to rely on Earth's internal magnetism. Yet the question of origin of Earth's magnetism is so natural that it is still an object of very competitive research. Nobody can ignore that the compass needle points toward the North, and it is a bit irritating that we still can't not offer a complete physical understanding of why it is so. The problem therefore becomes an active field of fundamental research in which significant progress has been made in the last few years using combined theoretical, experimental and a numerical approach. By its very nature, the problem is interdisciplinary and lies at the interface of physics, geophysics and applied mathematics. This problem has recently received considerable attention in the press because of concern of a possible reversal of polarity of the Earth's magnetic field in the near future. Considering that we risk seeing our planet unshielded from solar wind, understanding the field generation mechanism again appears to be a societal concern and a legitimate goal of environmental research.

THE ORIGIN OF EARTH'S MAGNETIC FIELD

When the Earth formed in 4.5 billion years ago, heavy elements concentrated at centre, as result 3000 km below our feet lies the largest of our planet's oceans: the core of liquid iron (mixed with traces of light elements), a sphere of 3400 km in radius. As the pressure increases towards to the heart of the Earth, the iron solidifies and we find a solid inner core, which occupies a volume of 1200 km of radius. It is the metallic core that the magnetic field of the Earth originates. Temperatures at such depth are above 3000 degrees K and thus well above Curies point (at which metal loses its ferromagnetic properties). A magnetic field can therefore only be sustained if electrical currents circulate in this ocean of liquid iron. However, Ohmic dissipation would suppress any unsustained electrical current in the Earth's core..."

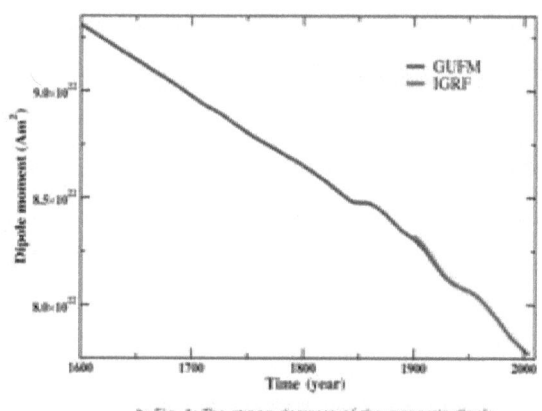

▶ Fig. 1: The strong decrease of the magnetic dipole moment over the last four centuries as revealed by the GUFM and the IGRF field models.

Figure 4/F-The strong decrease of the magnetic dipole moment

Unquote

About the same theme, we can't ignore the following that

cited from an academic literature of science in United Kingdom. At the first page we find right away:

Quote

> The Earth's magnetic field is generated in the fluid outer core by a self-exciting dynamo process. Electrical currents flowing in the slowly moving molten iron generate the magnetic field. In addition to sources in the Earth's core the magnetic field observable at the Earth's surface has sources in the crust and in the ionosphere and magnetosphere. The geomagnetic field varies on a range of scales and a description of these variations is now made, in the order low frequency to high frequency variations, in both the space and time domains. The final section describes how the Earth's magnetic field can be both a tool and a hazard to the modern world. First of all, however, methods of observing the magnetic field are described.

2 Geomagnetic field observations

2.1 Definitions

The geomagnetic field vector, **B**, is described by the orthogonal components X (northerly intensity) Y (easterly intensity) and Z (vertical intensity, positive downwards); total intensity F; horizontal intensity H; inclination (or dip) I (the angle between the horizontal plane and the field vector, measured positive downwards) and declination (or magnetic variation) D (the horizontal angle between true north and the field vector, measured positive eastwards). Declination, inclination and total intensity can be computed from the orthogonal components using the equations

$$D = \arctan \frac{Y}{X} \qquad I = \arctan \frac{Z}{H} \qquad F = \sqrt{H^2 + Z^2}$$

Unquote

We have gone through many theories but not the following combined theory which is most read at present time:

"The Earth and most of the planets in the Solar System, as well as the Sun and other stars, all generate magnetic fields through the motion of highly conductive

fluids. The Earth's field originates in its core. This is a region of iron alloys extending to about 3400 km (the radius of the Earth is 6370 km). It is divided into a solid inner core, with a radius of 1220 km, and a liquid outer core. The motion of the liquid in the outer core is driven by heat flow from the inner core, which is about 6,000 K (5,730 °C; 10,340 °F), to the core-mantle boundary, which is about 3,800 K (3,530 °C; 6,380 °F). The pattern of flow is organized by the rotation of the Earth and the presence of the solid inner core.

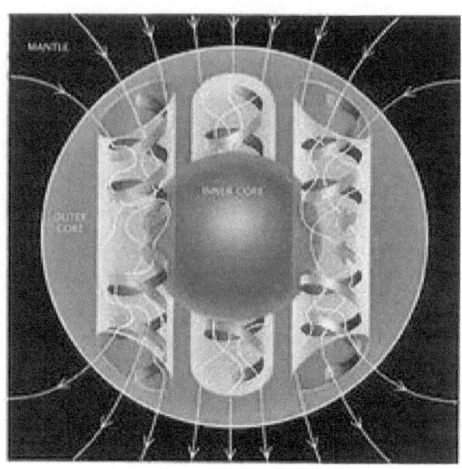

Figure 5/F-Rotors in the Earth

The mechanism by which the Earth generates a magnetic field is known as a dynamo. A magnetic field is generated by a feedback loop: current loops generate magnetic fields (Ampère's circuital law); a changing magnetic field generates an electric field (Faraday's law); and the electric and magnetic fields exert a force on the charges that are flowing in currents (the Lorentz force). These effects can be combined in a partial differential equation for the magnetic field called the magnetic induction equation:

$$\frac{\partial \mathbf{B}}{\partial t} = \eta \nabla^2 \mathbf{B} + \nabla \times (\mathbf{u} \times \mathbf{B})$$

...where u is the velocity of the fluid; B is the magnetic B-field; and $\eta=1/\sigma\mu$ is the magnetic diffusivity, a product of the electrical conductivity σ and the permeability μ. The term $\partial B/\partial t$ is the time

derivative of the field; ∇^2 is the Laplace operator and ∇ is the curl operator.

The first term on the right-hand side of the induction equation is a diffusion term. In a stationary fluid, the magnetic field declines and any concentrations of field spread out. If the Earth's dynamo shut off, the dipole part would disappear in a few tens of thousands of years.

In a perfect conductor ($\sigma=\infty$), there would be no diffusion. By Lenz's law, any change in the magnetic field would be immediately opposed by currents, so the flux through a given volume of fluid could not change. As the fluid moved, the magnetic field would go with it. The theorem describing this effect is called the frozen-in-field theorem. Even in a fluid with a finite conductivity, new field is generated by stretching field lines as the fluid moves in ways that deform it. This process could go on generating new field indefinitely, were it not that as the magnetic field increases in strength, it resists fluid motion..."

Although I am writing purely about external influence, I must make an unforgivable mistake if I don't mention these issues:

- The gap between Earth's rotation axis and magnetic pole: This issue will be re-compromised in the conclusion of this book. However, one of hypothesises about the origin of Earth Magnetism presumes that the "yolk" of the Earth, which induces the Earth Magnetism, used to be together with Earth's rotation axis but shocked by a huge meteorite then turns North side South. The "turn" has not been perfect and leaves the North Magnetic pole far from geometric North pole (about 11^0-17^0). By that author, the collision with the huge meteorite made a very serious disaster to the living in Dinosause Era, when all the Dinosause dead.

- The deviation and magnetic-longitude of "0" deviation.(Magnetic susceptibility of material).

The relevant documents may build a huge pile; we can't read every one of them but the following magazine- The GEOMAGNETISM-Magnetic Model and Navigation. The paper is to affirm that "the study of geomagnetism is one of oldest geophysical sciences. Geomagnetic fields have been observed and been used from ancient time." That paper also emphasizes that the magnetic compass needle does not point toward the North Magnetic Pole, but in the direction of the magnetic field line-horizontal composition- at the place where the compass is located.

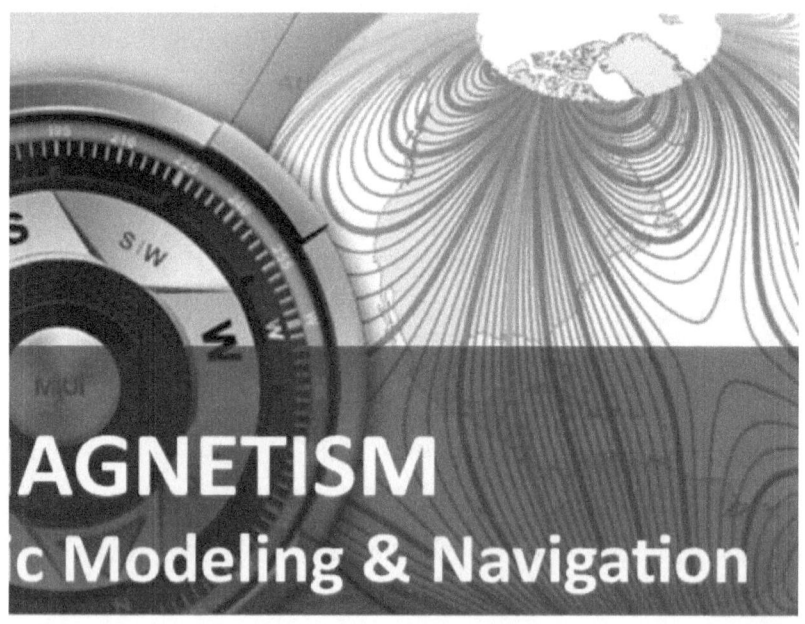

Image 6/F-Earth magnetism magazine

On the scope of International cooperation in research, the paper introduces World Magnetic Model (WMM), a joint product of the United States National Geospatial Intelligence Agency and United Kingdom Defence Geographic Centre. This model is used by many giant and important bodies such as U.S Department of Defence, NOAA, U.K ministry of defence, NATO, International Hydrographic Organization (IHO).

The paper is interesting and to be one of viable information sources for the author of this book.

The latest information from a science magazine is the model of some French scientists.

January 20, 2016 01:38pm ET

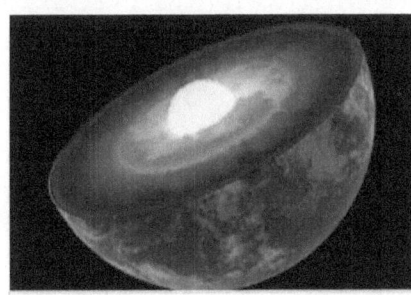

The power source for Earth's magnetic field may be magnesium that has been trapped in the core since our planet's violent birth, a new model suggests.

Magnesium is the fourth most common element in the Earth's outer layers, but previously, scientists thought there was almost no magnesium in the core. Iron and magnesium don't easily mix, and researchers thought that the Earth's core was mostly iron.

Image 7/F-Earth Magnetism produced in the core

The Earth core is supposed to be filled with molten

Magnesium since Earth's violent birth. The flow or move of the molten in Earth core is to produce Earth Magnetism.

Anyhow, those scientists must satisfy themselves as how the magnesium can be mixed together with iron, even if the molten magnesium is found in there.

Author's letter

Dear readers,

I have been a student of Vietnam Marine University, Navigation course L15 or 7411, from 1974 to 1979. At present, I am an independent researcher in Vietnam.

The first book written in English (in Y2K) is about management to sea-going crews, although no publishing is dispensed for public reader but many for colleagues' reference; since then I give up the "Business and human resource management" to concentrate in space, atmosphere and Earth science.

I actually return with Earth Science from 2007 but found no proper theme to study. Up to 2014, "The Deformed Earth" is found as an entry to return (posted on blog "ariveracity").

Like the last, this book is about the Earth. But unlike the last, this book is about magnetism of the Earth. This is specifically about contributions from atmosphere, Van Allen belt and Moon to the Earth Magnetism on top of the basic magnetic field created by rotors around the Earth core. The book is written in English with around 39000 words, 180 pages. It is completed on September 09th 2017 in Ha Noi, Vietnam.

On the occasion of this publishing, I would like to convey my great gratitude to my first university: The Maritime University of Vietnam, Hai Phong; where I am educated, encouraged to study and think about further theme in science. Meantime, I am obliged to every member of my family who facilitate and encourage me to research and write.

Especially, I must not forget the consultancy from The Professor, PhD Nguyen Dinh Noan and PhD Vu Bang who overcomes their age tiredness to check and gives advice to my work often.

I must apology for not writing in Vietnamese and hope to do so in future.

Thanks for your kind attention

Yours Faithfully,

Ha Noi September 09th 2017

Nguyen Van Cuong

CHAPTER II: AIR-EARTH CURRENT AND E.M

The exchange between air and earth surface is very complicated, and more than that there is even a very far difference between the process on land surface and one at sea surface. Both sea surface and land surface are huge themes to research; we do not discuss about the whole but something that relevant to the charges of land and sea surface.

I-General about atmosphere and its electrics:

1. Electric in air:

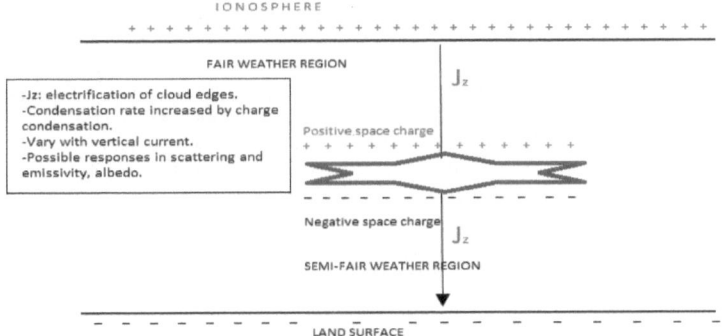

Figure 1/I-Atmosphere and electric in air (*From manual book*)

Atmosphere so far is recognized as the air with many layers surrounding the Earth. Certainly, atmosphere is the source of air for human being to breathe, nitrate for fertilizer, CO2 for plants and forest...The figures reposted in this chapter are definitely excerpted from manuals about atmospheric electric, which present and explain many complicated processes in our atmosphere.

EARTH'S PROTECTIVE SHIELD

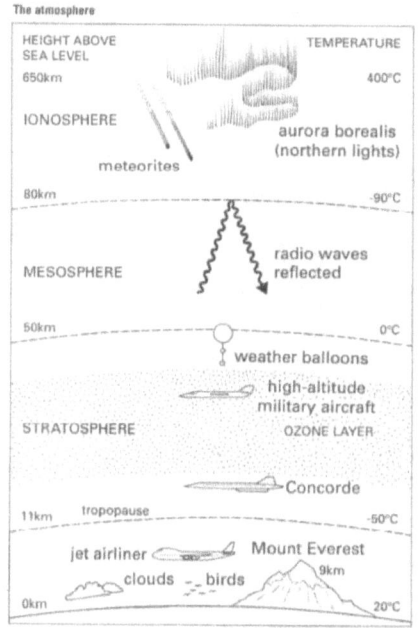

Earth is surrounded by invisible gases that form a thin protective blanket that we call the **atmosphere**. It contains the oxygen that we breath as well as other important gases such as nitrogen, carbon dioxide, water vapor, and ozone. It protects us from the Sun's harmful rays, burns and destroys meteors that are headed toward Earth, and it keeps our planet from having extreme temperature changes. Without this protective blanket we would have very hot days and freezing nights.

The atmosphere is divided vertically into four layers based on temperature: the *troposphere, stratosphere, mesosphere, and thermosphere*. Most of the atmosphere is in the troposphere which starts from ground level to about ten miles. We live in the troposphere were most of our weather conditions, like clouds, rain, and storms are produced in the troposphere.

Make as many words as you can from the word **ATMOSPHERE**
Note: You may find more than the lines allow

Figure 1a/I-The atmosphere

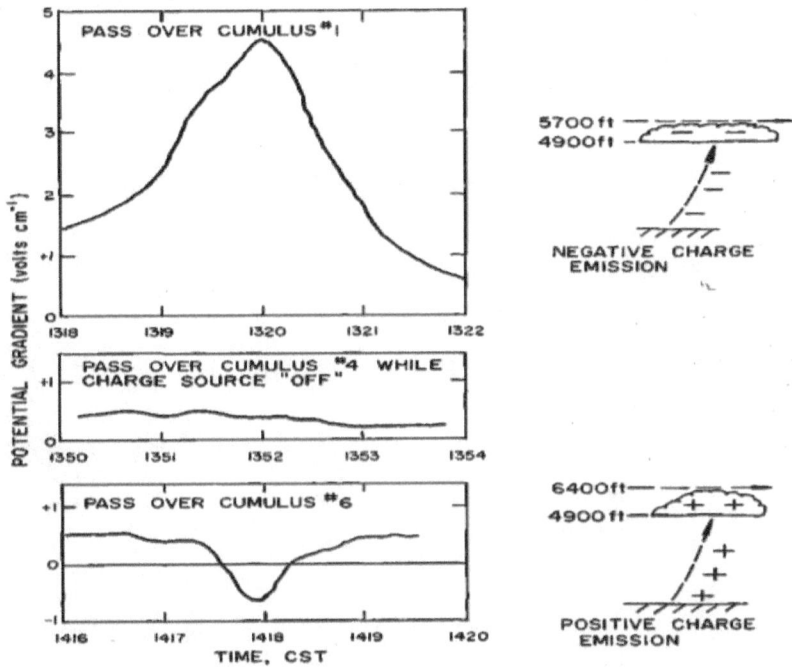

Figure 2/I-Air electric in atmosphere-The potential gradient (*From manual book*)

As atmosphere is a mixture of all kinds of air that come from soil and streams, rivers and oceans; therefore, atmosphere is very complicated. In addition, both soil and air are heated by sunlight and Earth core, so there must be many processes happening by which we can't anticipate. Our discussion in this book won't reach beyond the influence of atmosphere to Earth magnetism.

2-Convection: This is an activity that happens every time in atmosphere.

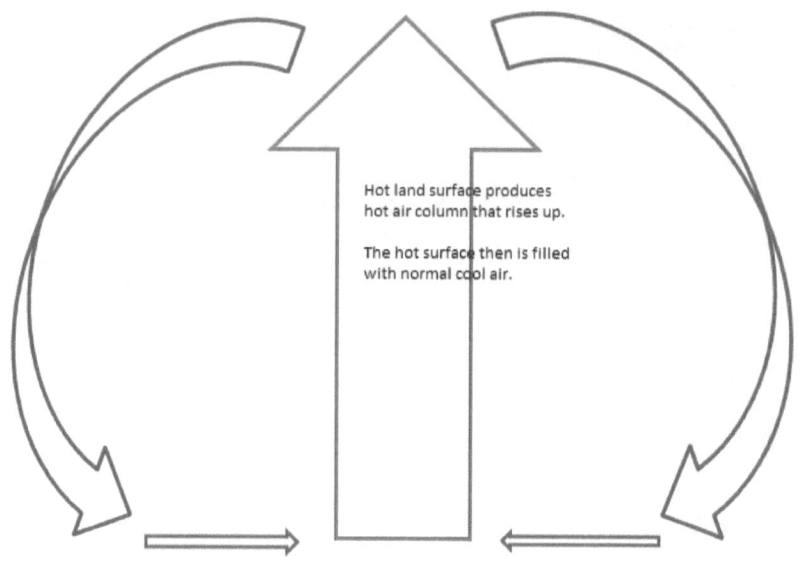

Figure 3/I- Convection illustration

a/ Tense and ease: The vital contrast is that the convection is rather tense on land and certainly ease on sea surface.

The cool air flocks to the low-pressure place where the hot air just rises; this process makes another story:

- During airflows touching Earth surface, the earth rotation energizes every of them un-equally on each different wind direction.

- As the Earth is rotating at all the time; every single air-flow is influenced, so each comes to fill a place but not a point; approximately each direction is tangent to an imaginable circle and not to eliminate any other.

- The closer to equator, the faster the Earth surface moves; this is how the Earth surface influences the wind, it is also named as Corriolis Effect; (Approximately 462.28 m/s at equator, 400.34 m/s at latitude $30^0N(S)$). This effect writes its own story in another issue-the storm or typhoon. We are expecting to discuss about it in another book.

- The solar heat does not distribute evenly everywhere, therefore some place is warmer than the others, and the convection is such different from one to another place, that's how complicated the terrestrial convection is.

Despite the convection does not contribute directly to E.M, it does make somewhere more positive or leave somewhere more negative, that's how it does contribute indirectly to E.M.

b/ Charge separation: In addition, no matter how much this issue is relevant to our problem, we can't ignore the magnetic force applying on either "+ion" or "–ion" during convection. The force obviously separates "+ion" from "–ion" which used to be mixed together in ion sheath on earth surface before convection. The separation is then writing its own story; we won't go all the way down every process in this book.

II-Air-earth current in general (J_{cd}):

Like any other global matter, air-earth conduction current is a huge issue that researched by several scientists. We don't go in advance of them but follow.

2/1. Concept:

There is not only one definition to this concept found, but in general the air-earth conduction current is so far recognized as a direct current of charges under the influence of air-earth electric field, and counted on square meter or square foot laid perpendicular to the general current.

2/2. Air-earth current research:

"The air-earth conduction current" is a theme that has been researched in centuries ago and the scientists still keep researching everywhere in the world. One of the pioneers in this science is C.T.R Wilson who pointed out that the global thunderstorm activity is energizing the earth-ionosphere electric circuit. Nonetheless we don't begin from C.T.R Wilson but with the following that cited from a magazine:

Measurement of atmospheric air-earth current density from a tropical station using improvised Wilson's plate antenna

C. P. Anil Kumar, C. Panneerselvam, K. U. Nair, K. Jeeva, C. Selvaraj, H. Johnson Jeyakumar, and S. Gurubaran

Equatorial Geophysical Research Laboratory, Indian Institute of Geomagnetism Krishnapuram, Tirunelveli-627 011, India

(Received July 28, 2008; Revised March 19, 2009; Accepted March 23, 2009; Online published August 31, 2009)

1. Introduction

Different classes of currents flow in the earth's atmosphere, and these can be classified according to the agency of transport. The conduction current constitutes the actual transport of charges under the influence of an electric field and is almost a direct current, while the displacement current is a fluctuating current that contains a spectrum of frequencies and does not involve any charge transport. The existence of the latter is due to the time variation of the electric field in the medium. This Maxwellian Displacement current density is given by,

$$J_d = \varepsilon_o \frac{\partial E}{\partial t} \quad (1)$$

where ε_o is the permittivity of free space, and $\partial E/\partial t$ is the time derivative of the electric field. For air-earth currents, the two main agencies that transport charge in the atmosphere are the electromotive force and the momentum of air. The current due to the electromotive force is the conduction current and is given by Ohm's law

$$J_{cd} = \sigma E \quad (2)$$

where σ is the conductivity, E is the electric field, and J_{cd} is the conduction current density. When the charge carriers are driven by air momentum, the current is called the convection current. Convection currents in the atmosphere can occur in different directions and intensities depending on the space charge density, air movement, the stability of the atmosphere, and gravity acting on charged particle suspensions.

Many attempts have been made in recent years to study atmospheric electrical parameters (Byrne *et al.*, 1993; Tammet *et al.*, 1996; Kar *et al.*, 2004). Long-term investigations agree well with the Carnegie curve (Ralph Markson, 1978), but short-term investigations reveal considerable deviations (Clayton and Polk, 1977). Dhanorkar *et al.* (1989) studied the variations in the conduction current during a solar eclipse, while Kamra *et al.* (1997) studied the effect of relative humidity. Datta and Bhattacharya (2004) shed light on the air-earth current during severe meteorological disturbances, while the effects of thermal power plant emissions on atmospheric parameters were studied by Monohar *et al.* (1989) and Monohar and Kandalgaonkar (1995). These studies determined that the site of GEC measurements must be free of atmospheric aerosols and convection activity. Otherwise they lead to noise that will obscure the weak signatures representing the global thunderstorm activity.

In the work presented here, the experimental site is 35 km east of the Bay of Bengal and 45 km from the Western Ghats. The landscape is nearly flat, and there are no trees in the vicinity of the sensors. Pollution is low, and scanty rainfall in this region enables a large number of atmospheric

---

Copyright © The Society of Geomagnetism and Earth, Planetary and Space Sciences (SGEPSS); The Seismological Society of Japan; The Volcanological Society of Japan; The Geodetic Society of Japan; The Japanese Society for Planetary Sciences; TERRAPUB.

2/3. What is the role of air-earth current in E.M?

For the matters relating to the sea surface, the book "Electromagnetic ocean effect" by NOAA (National Organization for Atmosphere Administration) and another "Oceanic Whitecaps and Their Role in Air-Sea Exchange Processes" by group of scientists have presented rather well. Those books are viable reference for the author of this book.

For the happenings on land, almost everything is presented in tens of books about "Atmosphere electric".

Our discussion is not about how to repeat or to re-check the works that done by those forerunners but how to perceive those findings and how the Earth Magnetism works under their light.

The following figures are displaying very important researching results that facilitate our research:

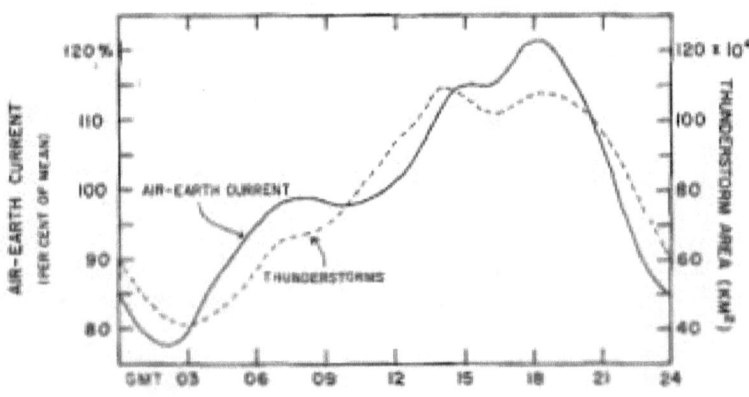

—Diurnal variation of the global thunderstorm area (Whipple [22]) and the fair-weather air-earth conduction current at Mauna Loa Observatory.

Figure 4/I-Thunderstorm area & air-earth conduction current-Mauna Loa.

The graph displays two plots that direct proportionally rise and fall against the same variable-the time. Although the graph doesn't say which from the two is pushing behind the other but the analogy in there does suggest the causal relation between them:

"The larger thunderstorm area expands, the more air-earth current increases".

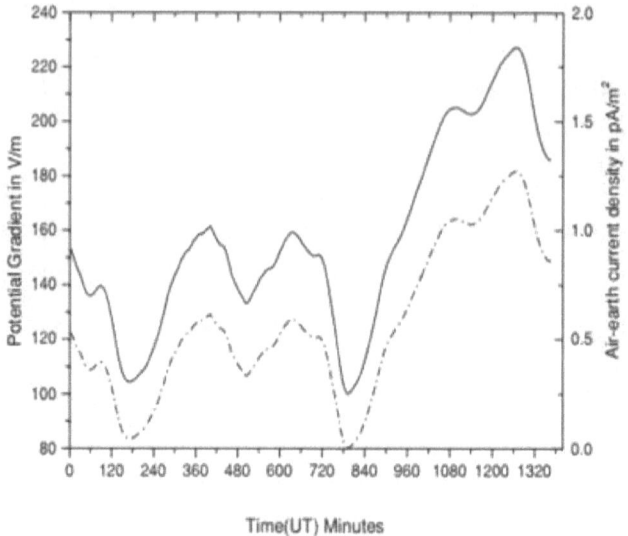

Plot of 1-min average of the air-earth current values of Wilson's plate (dotted line). The continuous plot shows the potential gradient values of passive antenna at measuring site Tirunelveli on the 25 April 2007.

Figure 5/I- Potential gradient & Air-Earth conduction current-Tirunelveli (India) at 8.73 N, 77.7 E. Recorded on 25th April, 2007.

The dotted plot displays air-earth current, average value in every minute with Wilson's plate. The continuous plot is for potential gradient with passive antenna.

Unlike the figure (5/II) that doesn't specify which is functioning on the other as variable. This graph (6/II) displays 2 in 1, where one of the two, potential gradient and the air-earth current density, definitely depends on the other:

"The steeper the potential gradient goes, the more air-earth current increases."

Although we are working with gradient but it is not in our discussion. Equation, differentiation as well as gradient are left for mathematicians; while the built graph is for us.

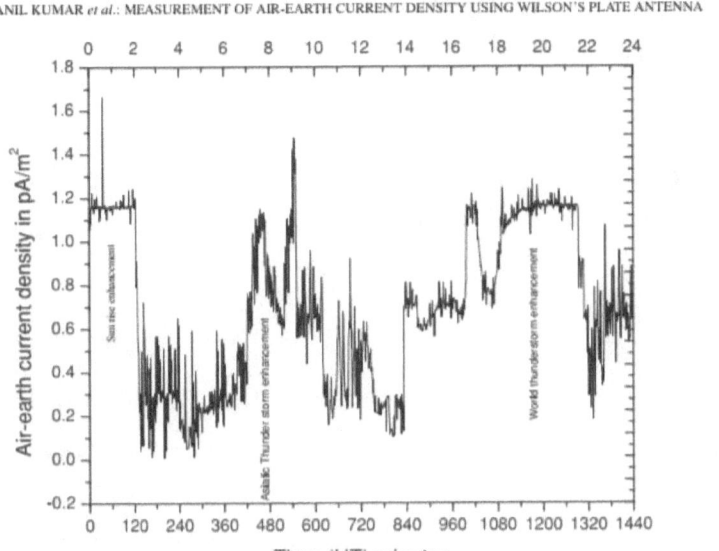

The figure depicts 1-min average observations of 25 March, 2007, showing the local sunrise effect as well as world thunder storm variations.

Fig.6/I-The true record at observer. (at 8.73 N, 77.7 E)

The above result is obtained at Tirunelveli (South India) on 25th March 2007, but actually, except the local sunrise enhancement, the result is quite analogous with every different observer in the world. (Moon time on the date: exactly on quarter or semi-moon, 11.42 rise, 18.16 hrs passing meridian, set on next day)

III-Scrutiny on Paradox, Questions and Causality:

Upon those records above and especially the true details of graph built on 25 Mar.2007 at Tirunelveli (South India).

Let's figure out the causal relation between sunshine and air-earth conduction current.

Argument: *(Although sea and land don't make a capacitor, in such a relation that the negative amplitude on one plate is responding to the positive amplitude on the other, but they are in electrically responsive relation).*

1-Assumptions:

a- When the Sun shines on the continent, it does not only energize the free ions on land surface, but also heats up the land to enhance the convection on there. The enhanced convection in its turn is to blow the air up to free the ion sheath from land surface.

b- The sea-land difference is then resulted in the real contrast between continent and ocean under sunshine. Land's free ions are energized under sunshine like the happening in photocell, and each locale is different to another; while all oceans are unified and make only single electrode due to their good electrical conductivity. So, land is localism while the sea is globalism.

c- There had been no solar storm happening to the atmosphere and Earth on 25 Mar. 2007, and so the U2 was unchanged within that 24-hour time of our problem.

Figure 7/I-Localism in land charge-one is different to another.

d-The Sun time at Suva (Fiji) indicates that there is no considerable difference between GMT+12 (as geographical calculations) and real time at longitude 180^0; therefore we may consider the meridian passing almost as instant local time at any locale in the world.

2-Arguments of Electric paradox & Earth surface charge: "losing electrons to be more negative?"

+a/ The continent is losing electrons and becomes more positive? This question is set to many people and definitely I don't believe the argument.

Like any material, continent is made of soil, stone and normal material. Under sunshine the continent must lose ion (both positive and negative) and become positive, this is a question?

+b/ The average ratio +2000/-1000:

In reality, the convection makes land surface losing ions of both positive and negative. The bold issue we consider is the ratio between positive ion and negative ion in every CC of air in land (not at sea side or sea surface), it is averaged as +2000/-1000, or (2) positive on (1) negative in city; so we should consider the land surface as losing +ions and definitely the Earth becomes more negative.

The charge of the Earth (continents and oceans) becomes more negative when larger part of continent is shined; this is a fact and the matter of fact. We rather discuss about its consequence.

Where no sunlight is shining on land, the surface is not heated up and so no convection is built and nothing special happens. As soon as sunlight shines on land, it is heated up and the convection is built and becoming tense. The convection clears up the land surface which used to be covered with ion sheath on it, and so the land surface becomes less positive or said specifically that the surface in convection becomes negative. We should emphasize the SURFACE and only the surface matter.

+c/ Balance charged status of the Earth:

Argument: The positive effect impacts on ocean becomes less, so to make the ion sheath (rather negative) above water surface loose, the ions in there can be freed rapidly; the charge balance is such maintained?

Not really, the reality does not happen in such normal way. The ions on sea surface are stacked on each other, accumulated and controlled by many different rules, not just static electric force. In short, the ease convection happens on there (ref. to "convection") and so the balance can be broken on favour of sea surface charge.

Thus, every ion freed from ocean surface would have to stay in the ion sheath; the lateness or "retardation" happens to the whole oceans to stabilize the Earth electric status.

The magic act is then working when Sunshine is more on land than sea; in the imbalance between positive land and negative sea charge, the sea surface seems to be unchanged while the land surface becomes more negative because losing positive ions.

Thus, the convection and retardation should be the motives to make to Earth more negative and even many more than that. This is still a question for further debate.

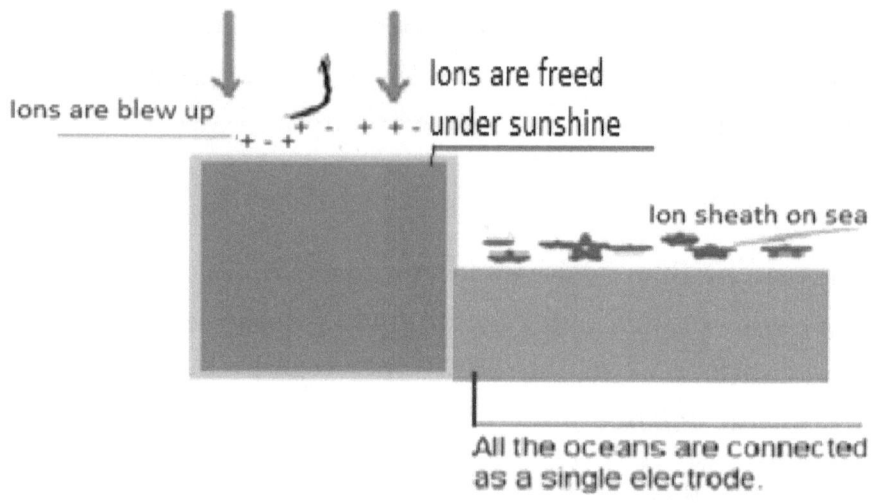

Figure 8/I-The ion sheath blew or retarded.

As mentioned above, the ion freed from land cum the difference in convection between sea and land are the key reasons of the rule "to lose ions to be more negative" and not the paradox that "to lose electron to be more negative". The paradox found happening to the set of Sea-Land-Sunshine on the Earth is wrongly ascribed to static electric force. Actually it is pure static electric, simple that losing positive ions is to be more negative.

This rule is almost applicable to whole the Earth and not specific to any locale or it is said: the globalism of the law.

From a book "The Oceanic Whitecaps and their roles in

Air-Sea exchange processes"-D.Reidel Publishing company, P.O. Box 17, 3300 AA Dordrecht, Holland; we may cite an excerpt of argument: *"It has long been recognized that as the Earth rotates, global thunderstorm frequency and the magnitude of the negative charge carried by the Earth are at maximum when the Sun shines on continents and convective activity is most tense, and at minimum when the Sun is shining on the ocean and convection is less tense."* (p219)

(PROBLEM: *(The real data are not sufficient to set up a pragmatic problem, therefore this is not solved in this book, just entertaining the physics and math lovers).*

Suppose the average condensation of ions and their ratio on world's land is (+1500/-1000) in every cubic centimetre (CC). Average land height (above average sea level):10 m. Average thickness of ion layer is 10 m for either sea or land.

Questions: What is the average ratio of +ions/-ions at sea surface to maintain an electrically neutral Earth? What is contribution of those to the Earth Magnetism if neutral Earth is maintained?(This problem is backing the argument presented in Biot-Savart expression that the magnetic intensity is inversely proportional to the squared distance from a charge to the centre of its hoop. In reality, everywhere the ion status is different and depending on local conditions).

The following is a researching record in some locale of Indian Ocean. We can have reference from the table, and note that it is not incorporated in the above problem.

Table 3: Periods and positions of ship at stationary positions and average values of small-, intermediate-, and large- ion concentrations (N_s, N_I and N_L respectively) of positive polarity and polar (Λ_+ and Λ_-) and total (Λ) conductivity during four cruises in the premonsoon, monsoon-onset and monsoon seasons. Values in parenthesis show their standard deviations.

Period	Location of ship (Stationary period)	N_S x 10^{-6} (m^{-3})	N_I x 10^{-6} (m^{-3})	N_L x 10^{-6} (m^{-3})	Λ_+ x 10^{-14} (Sm^{-1})	Λ_- x 10^{-14} (Sm^{-1})	Λ x 10^{-14} (Sm^{-1})
June 30- July 12, 2002 (Monsoon period -I)	16.9° N, 71.2° E	842 (±479)	3060 (±849)	6859 (±1980)	0.89 (±0.34)	0.83 (±0.17)	1.72
July 23– August 4, 2002 Monsoon Period -II)	15.4° N, 72.2° E	373 (±249)	1081 (±599)	11401 (±7267)	0.83 (±0.68)	0.65 (±0.27)	1.48
March 25- April 5, 2003 (Pre-monsoon period)	9.1° N, 74.5° E	895 (±557)	1116 (±1241)	3571 (±1549)	0.47 (±0.12)	0.54 (±0.23)	1.01
May 23- June 7, 2003 (Monsoon onset period)	9.1° N, 74.5° E	1515 (±741)	784 (±357)	6090 (±3510)	0.55 (±0.17)	0.51 (±0.20)	1.06

+d/ The Consequence and Causality: (Figure 8a/II).

We may make it clear by supposing that there is an absolute neutral point in figure 8a/II, which demonstrates a mock diagram of the Earth's potential: U1 and the other is U2- potential of ionized layer are taken.

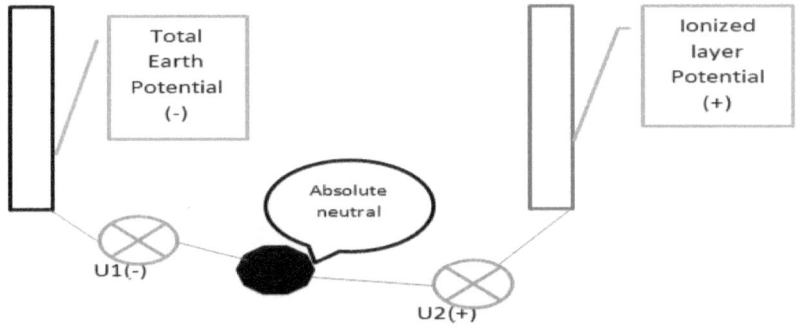

Figure 8a/I-Potential of ionized layer and Earth.

The potential U2 is relatively stable unless solar storm happens, while the magnitude of Earth charge-U1- is changing at all the time in accordance with the Sun shined land area. The more sunshine the more positive ion the land loses; while the negative sea surface is stable, this status makes the Earth more negative.

Consequently, the variation happens with the voltage between ionized layer and Earth:

$$V = U2 - U1 \qquad (1\text{-}3)$$

As soon as the potential U1 goes up-and-down, the voltage V goes down-and-up, or in another word:

The more negative Earth, the higher voltage between ionized layer and Earth is.

Certainly, any change in voltage leads to the change in air-earth conduction current. This is another way by which the Sun can influence the Earth; therefore:

With the brokerage of Earth's atmosphere the Sun can

affect the Earth by several ways.

+e/A question from critics: the Earth is losing ions at all the time and becomes un-neutral, how is the Earth balanced back?

This question is right and nice indeed. Actually, the chemical reaction is happening every time on the Earth to offset the any shortage. Moreover, another ion source is the atmosphere which is "paying" back in every lightning with a lot of ions of both positive and negative charges.

4-Review: Tirunelveli (South India) 25 Mar.2007.

- Sun time at Tirunelveli (South India): rise & set: 06.21 & 18.29 hrs local time. Solar noon: 12.25 hrs on 83^0 altitude.

- Every day, when the **GMT** points at 0.0 (or midnight), the local time at the longitude 180^0 is midday or solar noon by geographical time calculation.

The purpose of this arrangement is to match Sun's meridian passing time with local time throughout round clock or 24 hrs.

(To change all local time to GMT: Sun time at Tirunelveli: rise & set (Local time-5): 01.21 & 13.29 GMT. Solar noon: 07.25 hrs GMT.)

4/1-To match Sun meridian passing to graph:
We consider some specific locales under the Sun meridian passing from 0.0 minute to 1440 minute (24 GMT).

The graph begins at "0" minute to 1440 minutes while the Sun is at meridian above 180^0 to come.

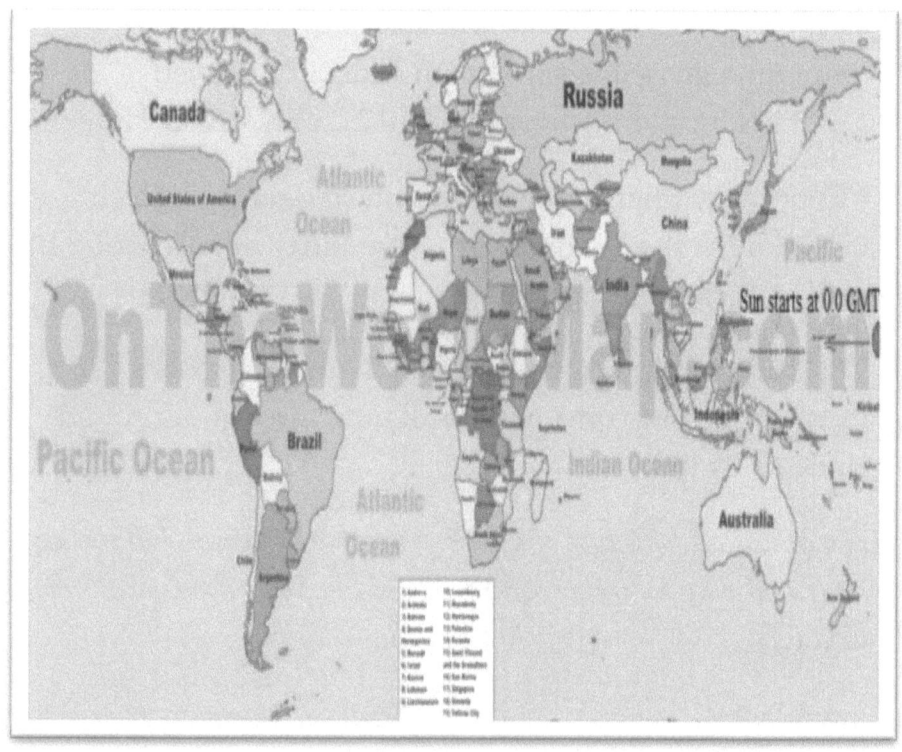

Fig. 9/I-Solar noon at 180^0E when GMT is 0.0

Therefore we got to turn the world map up-side-down and

attach to the graph to matching one to the other.

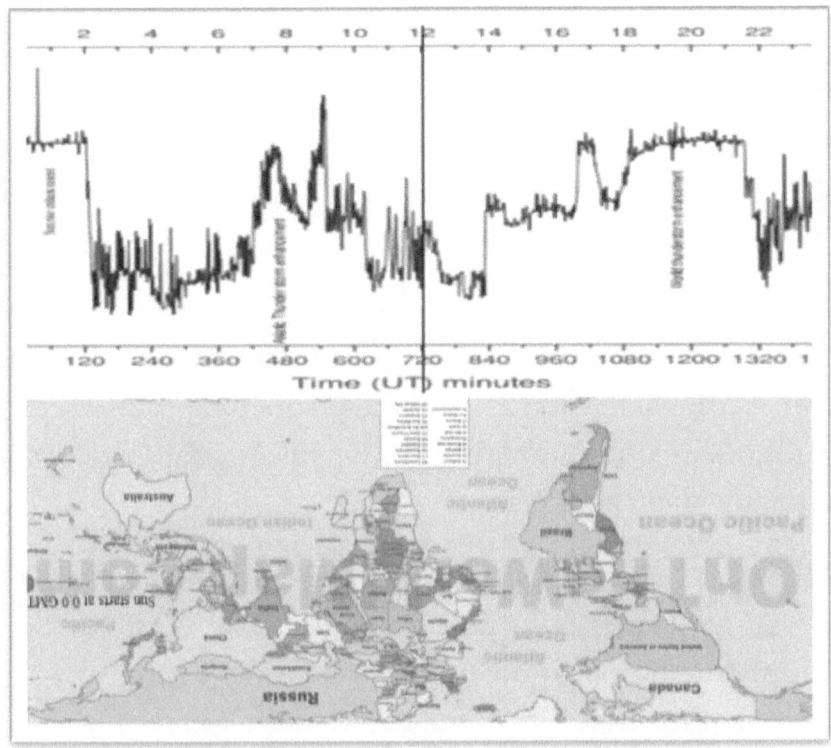

Figure 10/I-Two processes, one is approximately matching the other, from minute 00 up to minute 1440.

4/2-With a series of assumptions, we may carry out a review:

+Primary demonstration from graph:

- Let's start at 0.0 GMT from the longitude 180^0, the Sun is on meridian and begins its daily trip westward through a half of Pacific Ocean with East Russia, China, Australia,

New Zealand...South East Asia. It is still shining a bit on American continent but almost on Pacific Ocean; the South immense part of Pacific and a part of Asia, and it is expecting to come Australia. But the major shining area is on Pacific Ocean; therefore the image is demonstrating nothing tense except some lightning happens at Japan or Australia (as normally found).

- After 400 minutes, the Sun is above East Indian Ocean, about 2-3 hrs after leaving Japan, Australia behind. The air-earth current starts soaring up, this happening is attributed for thunder cloud establishing in Japan, Australia or even at Philippine and Indonesia.

- The next is Indian Ocean with Bangladesh, India, Pakistan...Middle East and East Africa. The West Africa and Atlantic Ocean and Europe & Africa ...Prime longitude Greenwich at minute 720th then switch to the other half of the trip from Greenwich (at minute 720th) to 180^0 (at minute 1440th).

- From minute 480 to minute 600, the Sun is almost above Africa. At this time, the Asian critical region is far away, while the thunder cloud in Africa is established.

4/3-<u>Comment</u>: In general, the air-earth conduction current varies against the following variables:

- The local soil characteristics: The above figure (10/II) hints that the soil of each area is showing off its characteristics under sunshine, this is likely the localism and still an unanswered question.

- The local thunderstorm area: The thunderstorm influences on air-earth current, this has been pointed out by Wilson in centuries ago. We don't debate again but depict it by our own way and it will work in the next chapter.

- Ocean electric conductivity: The globalism is feature of the ocean due to its electric conductibility. By such feature the potential at everywhere on the ocean, and seas connected with the ocean, is at the same value.

4/4-Special periods of time in a day:

It is known to everyone that approximately the Sun is shining 90^0 both westward and eastward longitudes from meridian, and so the meridian passing can justify the ratio of land shined area on ocean shined area but can't justify definitely the air-earth current which still depends on some others. Notwithstanding, we do consider the following special periods of time in a day for reference:

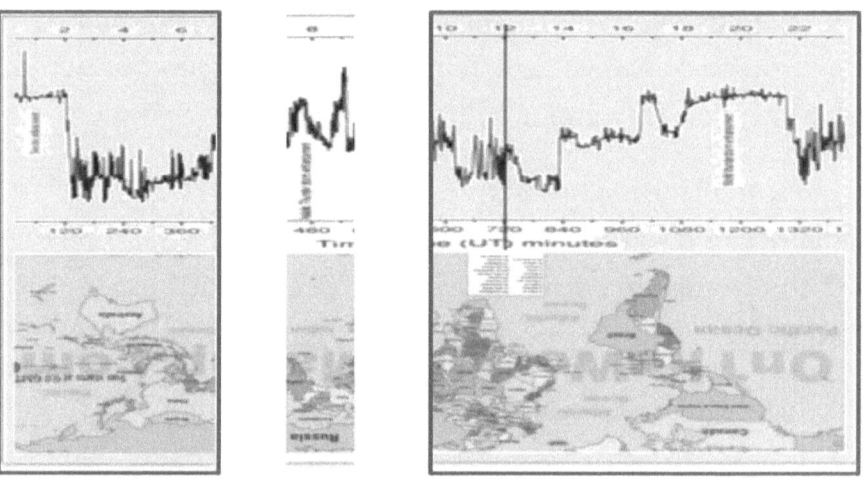

Figure 10a/I-Minute 420 until minute 560 the graph fluctuates and reaches max

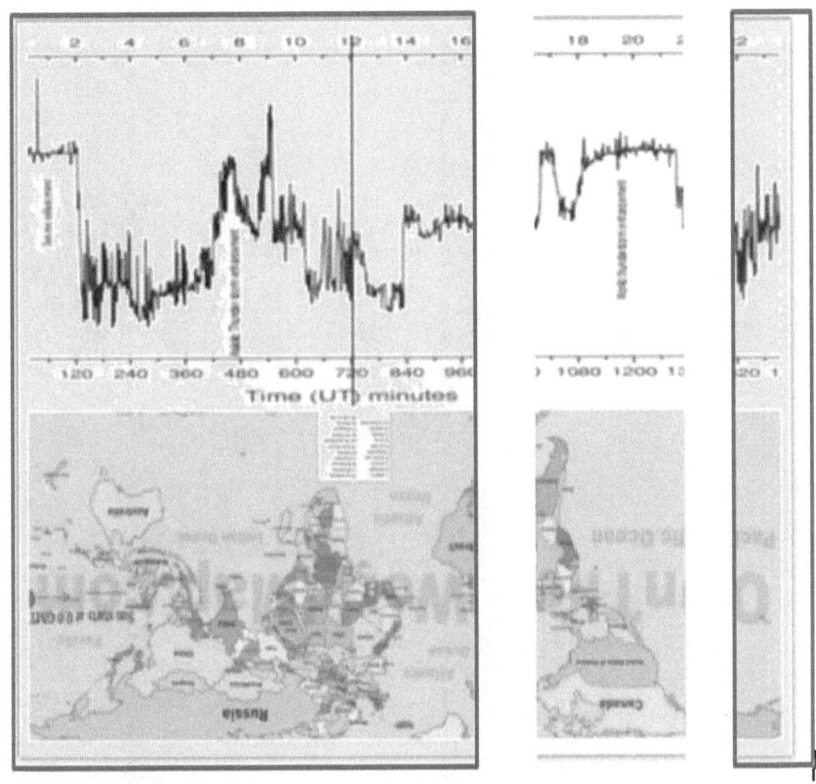

Figure 10b/I-From minute 920 until minute 1320 another top current in a day.

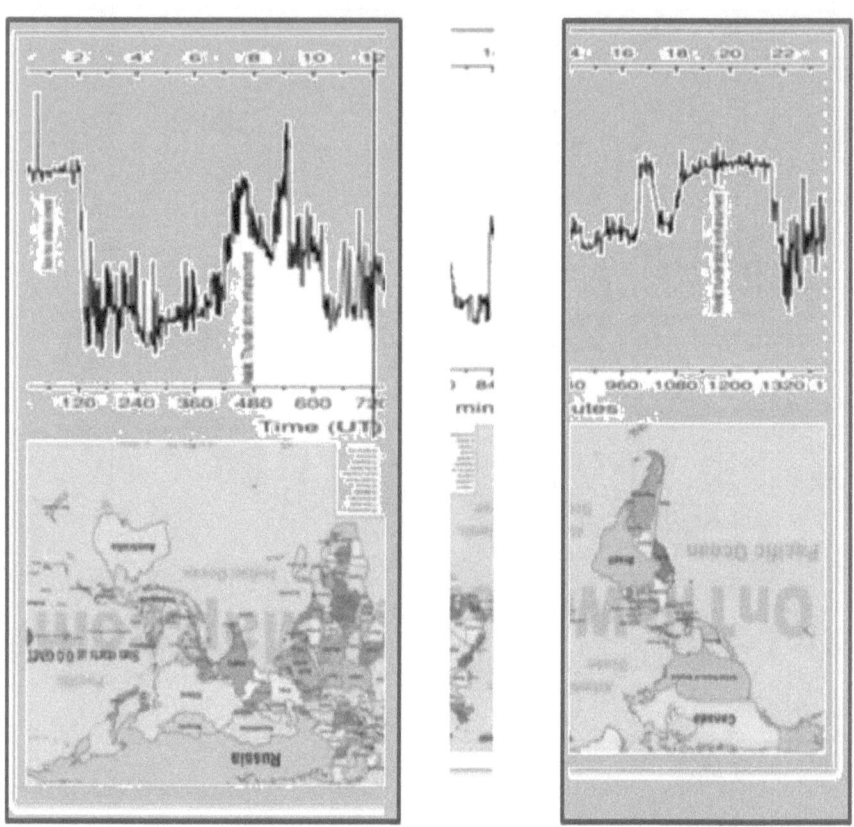

Figure 10c/I-Minute 780 until minute 840, the lowest current

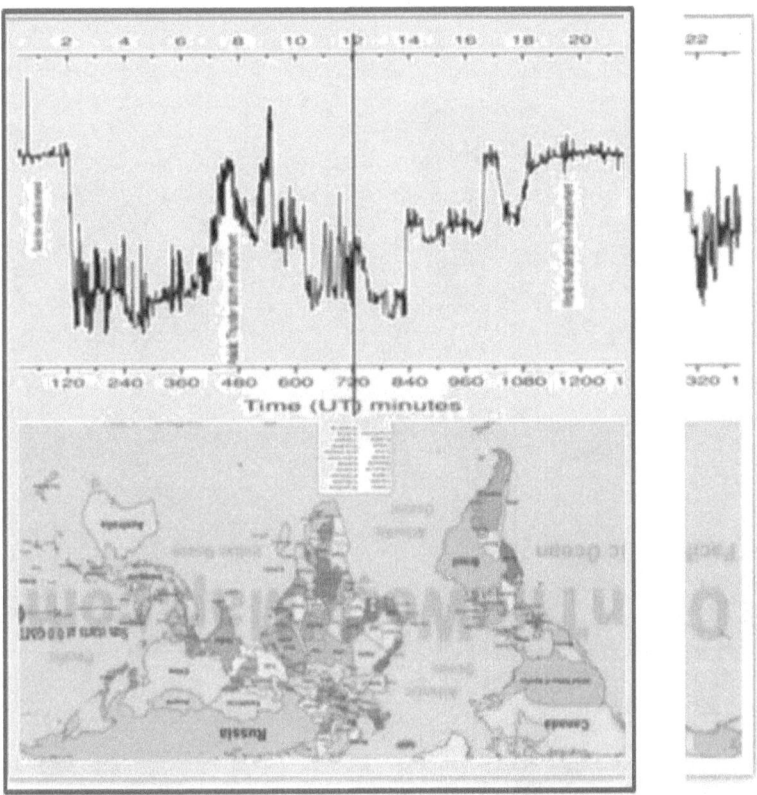

Figure 10d/I-From minute 1280 until minute 1440, another lowest current in a day.

After above demonstrations, we realize that the sunshine on land is a magic agent that makes the air-earth conduction current to vary. The following is another figure which demonstrates the hourly variableness of E.M with reference observer VSS:

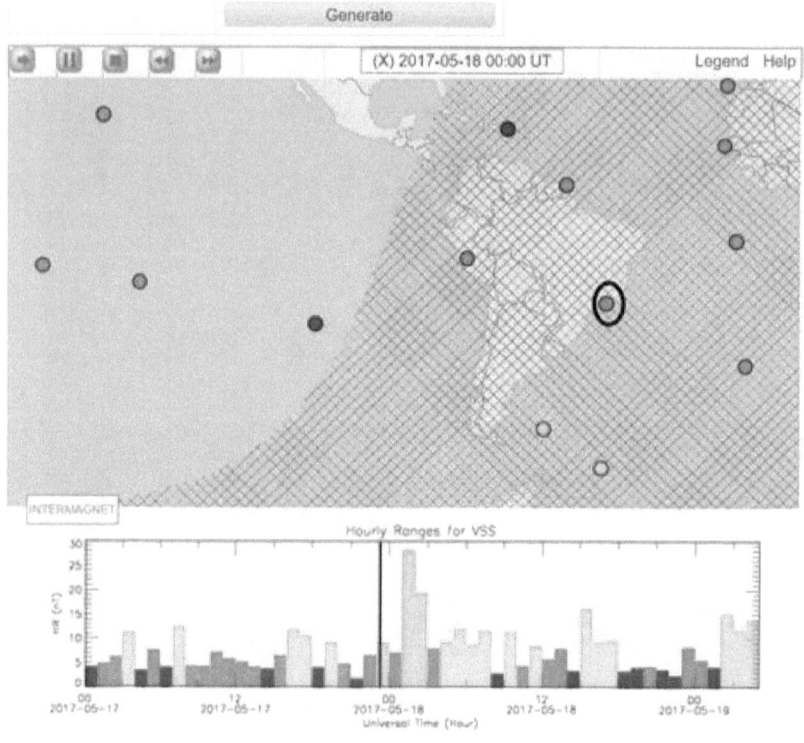

Figure 11/I-Hourly Range for VSS

We need discuss about the "how" at the hourly variation in some other chapter afterward.

CONCLUSIONS:

- By attaching air-earth current graph to up-side-down world map (figure 10/II), we realize that any variation in Sun shined land area must lead to respective change in the global total land charge.

- The more sunshine on land, the more positive ions it loses; then the more negative the Earth is. This issue is the key argument which was debated and recognized in history; we just recall in this chapter for reference.

- The air-earth conduction current incorporates with convection to make one local land different to another on term of electric charge.

- Land surface vs Sea surface: The convection on land makes land more negative; while there is less convection at sea surface.

- The particular characteristic of sea surface charge is globalism because all the oceans are connected and homogeneous electrically.

- For every Sun's position on the sky, we always find out a land area shined and the other is ocean area to be so; and the ratio between one to the other of those two under sunlight at each respective time is indirectly demonstrated in the figure 6/II as well as the figure 10/II -The Air-Earth conduction current.

- Although the indirect causality between air-earth conduction current and E.M is defined; the air-earth conduction current only contributes in some from 7 E.M compositions. However, we may have a look again at hourly E.M (Figure 11/II above) and speculate the contribution of air-earth conduction current in it.

CHAPTER III: PROBLEMS OF SEA VS LAND CHARGES

The last chapter presented a discussion about positively charged land surface and the negatively charged sea surface. They both are parts that constituted the Earth and moving together with Earth's rotation; if the charge of each is opposite to the other, then each induces magnetism that contrarian to the other. In the paragraph 2 (Argument of electric paradox) we even discussed and mentioned the average ratio +ions/-ions in every CC of air on land.

However, we leave that ion ratio for another discussion and work with graph of thunderstorm coverage for our problem in this chapter. So, we set and solve the following problems.

I-PROBLEM OF A SINGLE CHARGE ON THE EARTH SURFACE:

1-Problem:
- A charge of -100 Coulombs, locating at a fixed position P (30N, 100E) on surface a globe model of 100 m radius. Another charge +100 Coulombs, locating at the same fixed P on altitude h=10 m above surface of globe. How is magnetism induced by those 2 charges to rotation axis if its rotation is the same as Earth's cycle (24 hrs)?

a-Analyse:
This figure is displaying a global coordinate, where "O" is centre and P is a given point on global surface.

The latitude of P is also the angle "ϕ" between OP and equator's plane.

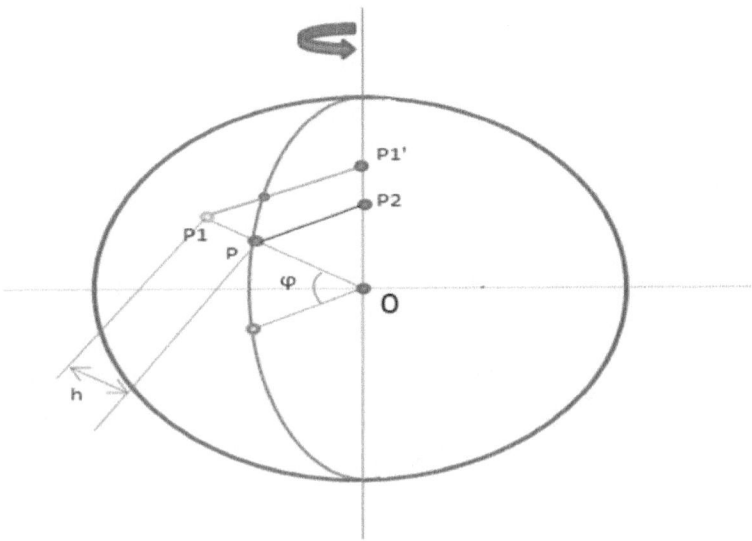

Figure 1/II-Charge at P versus P1 on global coordinates (not in scale).

The imaginary plane whereby the P moving on is the plane of latitude slice at 30N- depicted in the next figure.

The distance from P to global rotation axis is the value of "r" in our calculation:

$$PP2 = r = R * \cos(\phi)$$

The distance from P1 to global rotation axis is another value of "r_1" in our calculation:

$$P1P1' = r_1 = (R+10) * \cos(\phi).$$

The global rotation is to carry P and P1 together. In scope of electromagnetism, longitude of P (100 E) plays no role or helpless in our calculation.

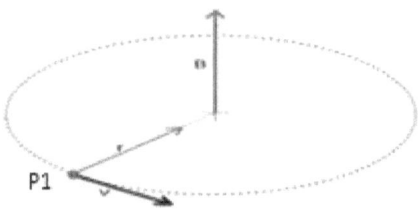

Figure 2/II-Slice of 30N. Vector $\vec{B}$ is product of 2 vectors $\vec{v} * \hat{r}$

The globe is rotating around its own axis at all the time. The "P" is fixed on globe's surface and moving along with global rotation. Every single charge moving around a certain point is to create a loop of electric current by which the magnetism is induced. The following is Biot-Savart expression (9.1.20) to facilitate our calculation:

$$\vec{B} = \frac{\mu}{4\pi} \frac{q\vec{v} \times \hat{r}}{r^2}$$

Permeability of the medium material between both P,P1 and globe rotation axis: As we are working with Earth model, therefore we suppose its permeability to be one of Earth, the average of Iron (99.95%) (Because Earth core is assumed as Iron), and normal air (μ_0 =4π*10^{-7} N/A^2). In reality μ is varying, the permeability at b-Assumptions: In order to solve this problem, we need some assumptions.

- The globe's average radius R=100 m so:

$r = R * \cos(\phi) = 100 * \cos(30^0) = 86.6$ m

$r_1 = 110 * \cos(30^0) = 95.26$ m

- Permeability of the medium material between P,P1 and globe rotation axis: As we are working with Earth model, therefore we suppose the permeability as one of Earth, the average of Iron (99.95%) (Because Earth core is assumed as Iron), and normal air ($\mu_0 = 4\pi*10^{-7}$ N/A²). So we have:

$\mu = 1/2 \ (0.25+4\pi*10^{-7}) = 0.12500126$ N/A²

General formula for speed (v) of P: $v=2\pi r/24*60*60$ m/s, replace "v" to Biot-Savart expression:

$$B = \frac{\mu}{4\pi} * \frac{q*v}{r*r} \quad \text{And so dB} = \frac{0.12500126}{4\pi} * \frac{q*2\pi r}{r*r*24*60*60} =$$

$$dB = 0.12500126*q/(r*48*60*60) =$$

$dB = 7.233869*10^{-7}*q/r$ (named as I-Biot-Savart).

Although the coefficient $7.233869*10^{-7}$ is derived from μ, π and cycle (24 hrs), it is interpolated from Biot-Savart expression, so we name it as "interpolated Biot-Savart" or I-Biot-Savart for further reference.

c-Solution with Biot-Savart:

Solution 1:
- Given charge at P: q = -100 Coulombs

- Radius of the loop: r = 86.6 m

- Product of vector speed "v" and unit vector "r" is a vector that parallel with Earth's rotation axis, pointing to true North and valued |v|, because vector $\vec{v}$ is perpendicular to unit vector $\hat{r}$.

- Value of Δb1 is calculated step by step with the above Biot-Savart interpolated as: $=7.233869*10^{-7}*q/r=-7.233869*10^{-7}*100/86.6=-8.353197*10^{-7}$

Solution 2:
- Given charge at P1: q1 = +100 Coulombs

- Loop radius: 95.26m

- Value of Δb2 is also calculated step by step with the above Biot-Savart interpolated as:

$= 7.233869*10^{-7}*q/r1 = 7.233869*10^{-7}*100/95.26 =$

$= 7.593816*10^{-7}$

d-Summary words: (apply I-Biot-Savart).

$$\Delta 1 + \Delta 2 = 7.233869*10^{-7}(\frac{q1}{r1} + \frac{q2}{r2}) \text{ therefore:}$$

$-8.353197*10^{-7} + 7.593816*10^{-7} =$

= <u>-0.7593816*10⁻⁷ Tesla</u>

If two oppositely charged points of the same absolute values are placed on the same position at 2 different altitudes; resulting the distance from each to the rotation axis to be different to the other.

The intensity of magnetism contributed by each charge is different to the other. The closer to axis, the more it contributes; that's rule. The total of the two is in favour of the inner charge.

II-PROBLEM OF OCEANS VS CONTINENTS

1-Basic argument:

The Sun is rather far from Earth, so it can't apply any magnetic influence on the Earth directly but indirectly through brokerage of atmosphere, continent and ocean surfaces. Beyond the influence that almost considered in the last chapters, this is a problem for the contributions of continent and ocean (or sea & land as I name them in this book) to the Earth Magnetism in the light of Biot-Savart laws.

2-Thunderstorm matter:

From atmospheric research, the thunderstorm area varies against the time with 24-hr revolution or daily cycle. The following is a graph that re-built by the data taken at Mauna Loa Observatory (Hawaii Islands):

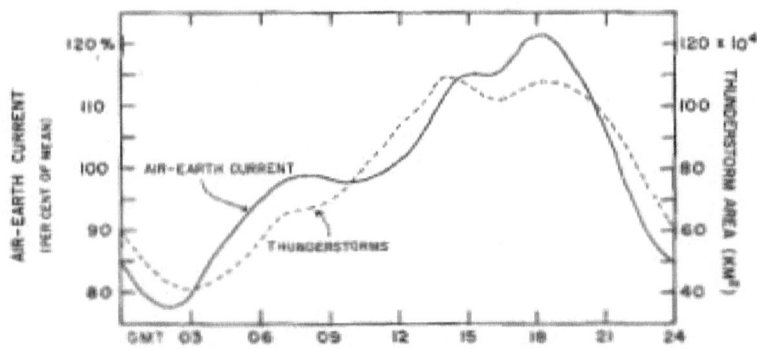

—Diurnal variation of the global thunderstorm area (Whipple [22]) and the fair-weather air-earth conduction current at Mauna Loa Observatory.

Figure 3/II-This is observation result from Mauna Loa.

The graph is indicating that the Americas continent is the place of thunderstorm and lightning, while the less thunderstorm continent is Asia. The average thunderstorm continent is Africa which is at the same peak as Americas.

The above figure is divided approximately into 3 trapezia and 1 rectangle to calculate the total area of thunderstorm and its average within 24 hrs. The result is the average area under daily thunderstorm "$80*10^4$ km^2", used as entry for our problem.

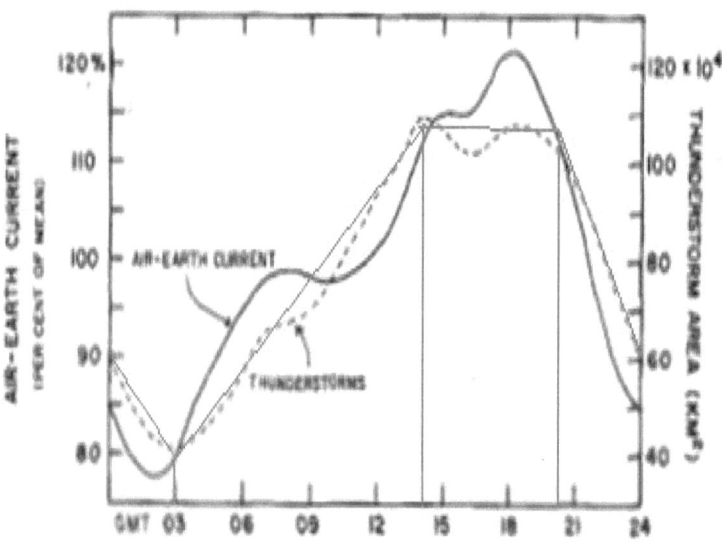

—Diurnal variation of the global thunderstorm area (Whipple [22]) and the fair-weather air-earth conduction current at Mauna Loa Observatory.

Figure 4/II-Global average area under thunderstorm daily.

Thunderstorm and lightning on 30 March 2017 (as instant only, the situation is changing at all the time).

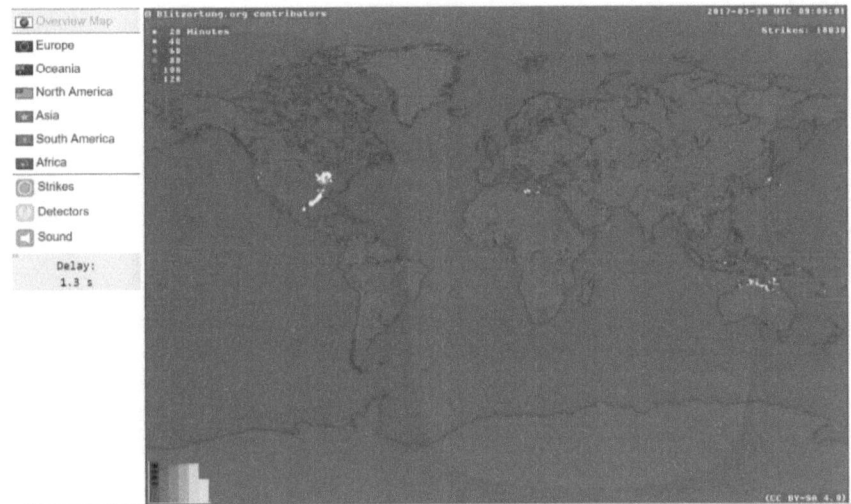

Picture 5/II-An instant image (courtesy from Blitzortung.org)

The world thunderstorm map is indicating that the thunderstorm is covering almost within 55N to 55S, where some spotted lands should be mentioned in the South semi-sphere such as Australia, Indonesia. So, we consider a reasonable way to distribute the area of $80*10^4$ km^2 under daily thunderstorm, make those figures the entries to our problem.

Note: (*Under the thunderstorm, the area is almost positively charged and assumed to be so*)

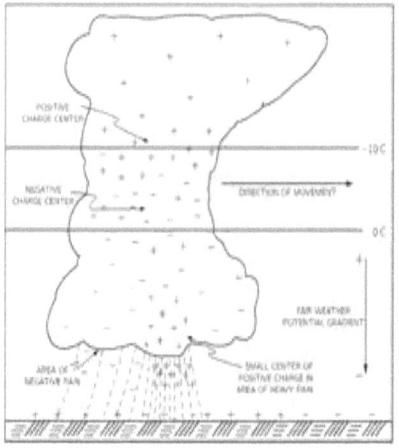

The distribution of electrical charges in a mature thunderstorm cell. [From U.S. Department of Commerce Weather Bureau Report, June 1949.]

Figure 6/II-Illustrative structure of a thunderstorm cell

(The above figure is re-built after many real flights through a super cell, the pilots who performed the flight surveys were recognised as braves in atmosphere research.

The obvious sign illustrated above is the positive land surface that expands as large as the thunderstorm cell is. The positive patch on land, as matter of fact, must be laid there until all the charges above to be discharged.)

3-Problem:

Daily global average area under thunderstorm is $80*10^4$ km². If the land under thunderstorm is positively charged (supposed average +1000 Coulombs/km²); reciprocally the oceans are negatively charged so that they can balance with the land positive charge (P65-Bal.).

Assume that the land average height is 10 m above mean sea level, while sea level is assumed as at its average level during our calculation.

What is the contribution of sea and land to Earth's Magnetism?

X
X   X

There are obviously several ways to solve the problem. Our way is to attribute a certain charge to each potential location, calculate contribution of each and totalize at the end of this section.

The seas (and oceans) are considered as one single electrode because salt water in ocean is homogenous with high electric conductivity. The assumed value of electric charge is distributed evenly to everywhere of seas and oceans. We will calculate the ocean contribution to earth's magnetism and summed up with contribution from lands.

3a-Solution-Land contribution ($\Delta 1$):

<u>Expression</u> I-Biot-Savart (p84) for $\Delta(i)$:

$\Delta(i) = 7.233869 \times 10^{-7} \times q(i)/r(i)$

Back to our problem, it should be simplified by distributing approximately the total given land charge to 6 different places; each of them represents a specific locale where the thunderstorm land is dominating its region.

The land charge is estimated on basis of daily air-earth record (figure 4/II), and every charge is positive, furthermore as noted before that each diminishes the E.M.

From entry, we have $80*10^4$ km² area and assumed 10^3 Coulombs/km²; so total charge on land is assumed as $8*10^8$ Coulombs. From every position, we always find the distance to Earth rotation axis (assume that average height of land charge is 10 m above average sea water level).

With approximately estimated percentage of areas under thunderstorm, we also attribute average area under thunderstorm for each different location. Eventually an assumed charge at each location can be found. A table of relative calculation is established as following:

- **Assumption1:** by percentage, the areas under thunderstorm with respective charges are approximately divided as following:

+ America continental (AC): 20 pcts, (35N, 90W.)

With $r(1) + 10 = 5,207,195.7$m, and $q(1) = 16*10^7$

+ North Africa & Europe (A&E): 20 pcts, (40N, 20E.)

With $r(2) + 10 = 4,869,601.3$m, and $q(2) = 16*10^7$

+ South Africa (S.A): 10 pcts, (10S, 30E.)

With $r(3) + 10 = 6,260,235.9$m, and $q(3) = 08*10^7$

Table 1 - Δ(1)

Position	Latitude	"r+10" Distance to R. Axis (m)	Area under T. Storm (%)	Charge	Δ(ci)Mag. Contribution Δ(ci)
AC	35N	5,207,195.7	20	$16*10^7$	$2.22273006*10^{-5}$
A&E	40N	4,869,601.3	20	$16*10^7$	$2.37682505*10^{-5}$
S.A	10S	6,260,235.9	10	$08*10^7$	$0.92442127*10^{-5}$
Jap.	35N	5,207,195.7	15	$12*10^7$	$1.66704754*10^{-5}$
P&I	0.0	6,356,810.0	15	$12*10^7$	$1.36556587*10^{-5}$
Aus.	25S	5,761,227.3	20	$16*10^7$	$2.00879996*10^{-5}$
Total magnetism contributed by land charge Δ(1) in Tesla					$10.56539*10^{-5}$

+ Japan and its' neighbour (Jap.): 15 pcts., (35N, 140E.)

With r(4) + 10 = 5,207,195.7m, and q(4) = $12*10^7$

+ Philippine and Indonesia (P&I): 15 pcts., (0.0, 115 E.)

With r(5) + 10 = 6,356,810.0, and q(5) = $12*10^7$

+ Australia (Aus.): 20 pcts. (25 S, 140 E.)

With r(6) = 5,761,227.3 m, and q(6) = $16*10^7$

(Thunder cloud (negative) makes its opposite patch on land positive (see the super thunderstorm cell), each lightning discharges a lot of charge on both cloud and land charged patch. This is assumed as one of motivation to the surge of E.M every day. Therefore, we think about the following problem.)

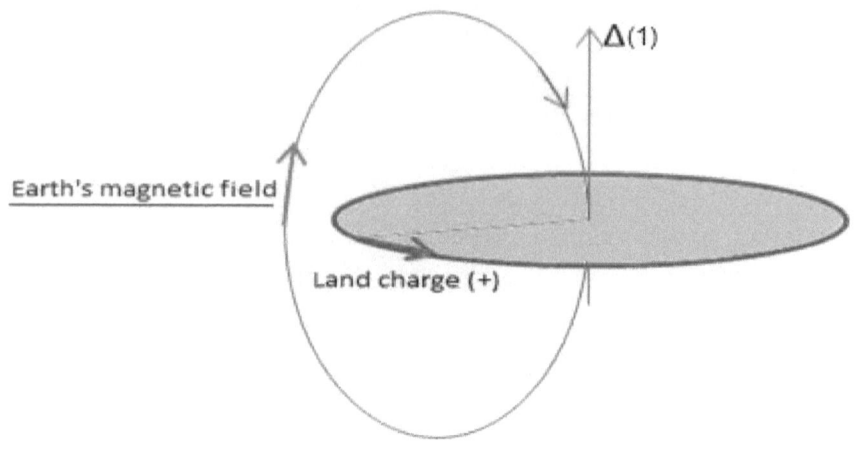

Figure 6b/II-Land charge (+) contribution-$\Delta(1)$

Problem 1:

A thunder cloud patch of following properties:

- Coverage: 10 km²

- Average negative cloud height: 2000 m

- Location: 30N, 90W as coverage's centre.

- Total charge: 10^6 Coulombs.

There is no more charge on cloud left, or the total 10^6 Coulombs will disappear after the lightning.

Question 1: if we consider the contribution of thunder cloud to E.M, what are the contributions of both cloud and its opposite patch on land to E.M?

Question 2: How is the surge in E.M caused by the lightning?)

This problem is not solved or discussed in this book.

3b-<u>Solution-Ocean contributions ($\Delta 2$)</u>: We call back I-Biot-Savart.

$\Delta(i) = 7.233869 * 10^{-7} * q(i) / r(i)$

From ocean facts, the four oceans are covering 70% of global area. Every ocean is connected to the others, the water in ocean is salted and always at high electric conductivity; therefore sea water is considered as homogenous as a single electrode in relation to any other object.

As assumption in the entry of this problem, our globe is assumed as a neutral object, so the total electric charge is "nil". Because the land is charged as counter part of thunder cloud, so the oceans must have charge that opposite to land for neutralizing the land charges. By this argument, we consider the total sea-charge equal to total land-charge and evenly distributed on every ocean and seas connected to them.

Figure 7/II-Globe with Longitude-Latitude network

+ From the air-earth current research, our calculation gives an approximate result for total land charge under thunderstorm +80*10⁴*1000 Coulombs. Therefore:

Total Sea Surface Charge (SSC stands for them from now onward) should be as much as -80*10⁴*1000 Coulombs or:

= -8*10⁸ Coulombs

We assumed that the total is maintained and distributed evenly over every ocean throughout our calculation.

+ From the figure 7/II above, we divide the globe into 6 global chunks (horizontal bands from North down to South). Each chunk expands 30^0 latitude, we calculate "r" (distance to Earth's rotation axis) from average latitude of the band such as 15 is average of 0-30; 45 is average of 30-60; 75 is average of 60-90.

After the distance "r" calculation, each band can be considered as a rim or a hoop by which the negative charge locates on. The figures are approximately assumed as following:

3/2/1-From 60N to North Pole including Artic, $20*10^6$ km²

Figure 8/II-Arctic and its substitute

- Distance to axis r(1)=Re*cos(75)=1,645,216 m

- Water surface area: 19,889,355 km²

Figure 9/II-Band of 30N-60N latitude degrees and its substitute.

- Distance r(2) = Re*Cos(45) = 4,494,936.4 m

- Water surface area: 44,745,958 km²

Figure 10/II - Band of 0-30N latitude degrees and its substitute

Distance to rotation axis: r(3) = Re*Cos(15) = 6,140,197 m

Water surface area: 74,576,252 km²

Figure 11/II-Band of 0-30S latitude degrees and its substitute

Distance to rotation axis: r(4) = r(3) = Re*Cos(15) = 6,140,197 m

Water surface area: 119,322,568 km2

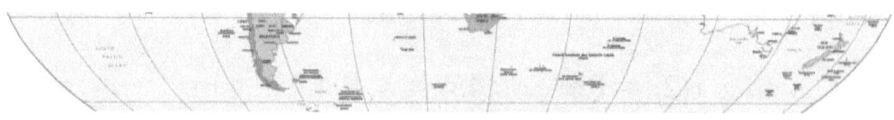

Figure 12/II- Band of 30-60S latitude degrees and its substitute

Distance to rotation axis: r(5) = r(2)Re*Cos(45) = 4,494,936.4 m

Water surface area: 80,542,806 km2

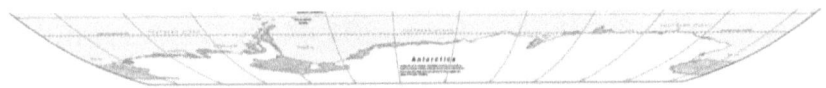

Figure 13/II- Antarctic and its substitute

- Distance to axis r(6) = r(1) = Re*cos(75) = 1,645,216 m

- Water surface area: 14,915,671 km^2

XXXXX

We are dividing the globe into 6 chunks or bands, each is far from Earth's rotation axis at an average distance, denoted as r(i) where **"i"** is varying from 1 to 6. We also attribute approximately a certain sea surface area to each chunk in condition that total charge can balance $+8*10^8$ Coulombs of all continent summed up.

We have calculated Δ(1) the land contribution to E.M, now we do the same to find out sea surface contribution Δ(2) to E.M. Our job at this stage is to figure out how the 6 chunks contribute to E.M.

Solution:

Total area: 352,876,000 km2,

Charge per s.km: $-8*10^8/352{,}876{,}000 = -2.267085$ Coulombs/km²

We have every q(i) for respective chunk as following:

q(1) = -2.267085*19,889,355 = -45,090,858 Coulombs

q(2) = -2.267085*44,745,958 = -101,442,890 Coulombs

q(3) = -2.267085*74,576,252 = -169,070,702 Coulombs

q(4) = -2.267085*119,322,568 = -270,514,404 Coulombs

q(5) = -2.267085*80,542,806 = -182,597,387 Coulombs

q(6) = -2.267085*14,915,671 = -33,815,094 Coulombs

We again do apply I-Biot-Savart or Biot-Savart 9.1.20 to each chunk from North to South:

$\Delta(i) = 7.233869*10^{-7}*q(i)/r(i)$:

- With r(1) = 1,645,216 m, q(1) = -45,090,858 Coulombs

  $\Delta(w1) = -1.982605*10^{-5}$

- With r(2) = 4,494,936 m , q(2) = -101,442,890 Coulombs

  $\Delta(w2) = -1.632558*10^{-5}$

- With r(3) = 6,140,197 m , q(3) = -169,070,702 Coulombs

  $\Delta(w3) = -1.99185*10^{-5}$

- With r(4) = 6,140,197 m, q(4) = -270,514,404 Coulombs

  $\Delta(w4) = -3.1869755*10^{-5}$

- With r(5) = 4,494,936 m, q(5) = -182,597,387 Coulombs

Δ(w5) = -2.938608*10^{-5}

-With r(6) = 1,645,216 m, q(6) = -33,815,094 Coulombs

Δ(w6) = -1.48682*10^{-5}

Table 2-Δ2

Chunk No.	Latitudes	Distance to rotating axis (r)	Charges	Magnet contribution $\Delta_{(wi)}$
1	60-90 North	1,645,216 m	-45,090,858.	-1.982605*10^{-5}
2	30-60 North	4,494,936 m	-101,442,890	-1.632558*10^{-5}
3	00-30 North	6,140,197 m	-169,070,702	-1.99185*10^{-5}
4	00-30 South	6,140,197 m	-270,514,404	-3.186975*10^{-5}
5	30-60 South	4,494,936 m	-182,597,387	-2.938608*10^{-5}
6	60-90 South	1,645,216 m	-33,815,094	-1.48682*10^{-5}
Total contribution of sea surface charge (SSC), Δ2				-13.219416*10^{-5}
Total contribution of land charges (Δ1)forwarded from table 1				+10.56539*10^{-5}
Average sea-land contribution to E.M				-2.654026*10^{-5} Tesla

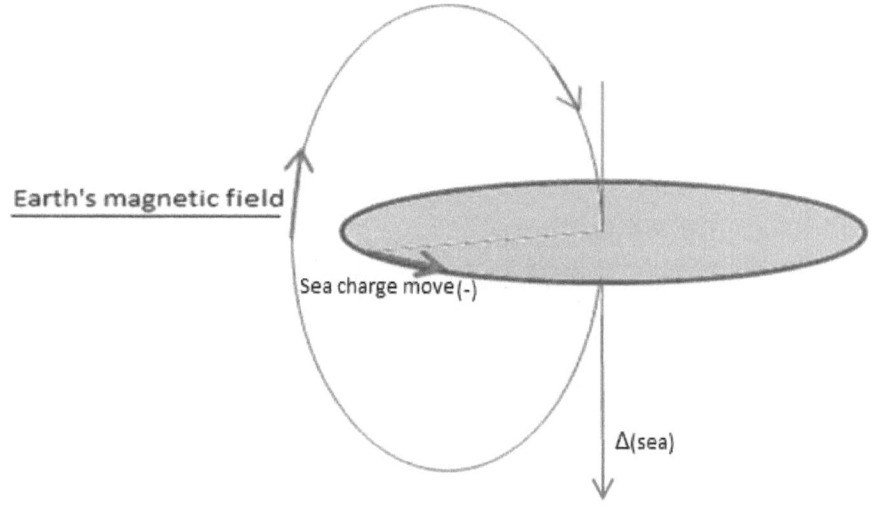

Figure 14/II-Sea Surface contribution to E.M

CONCLUSIONS (for part I):

- In addition to underground rotors, the land charges (figure 6b/II) and sea surface charge (SSC), (figure 14/II) contribute to the E.M. Although the land charges as well as sea surface charge are belonging to the Earth but totally influenced by external variableness regardless how we denote it.

- Although those two problems are set and solved with assumed figures, the results are demonstrating that:

+ The total of $\Delta(1)+\Delta(2)$ is on favour of $\Delta(2)$, indicating that the influence of the sea is more than that of the land, and supporting E.M or adding on it.

+ The lightning producing jolt: lightning frees the charges for both thunder cloud and its opposite patch on land; so as consequence of lightning, the $\Delta(1)$ must drop right away and considerably. The drop of $\Delta 1$ produces a huge magnetic jolt (total $\Delta(1)+\Delta(2)$ soars due to $\Delta(1)$ drop) or the magnetic surge, demonstrated in daily E.M record.

- The divided Earth: might be divided by many different ways, the sum of $\Delta(1)+\Delta(2)$ might be changed because of such different dividing; but the ultimate demonstration is unchanged and that: The negative charge on sea surface stands up to the positive charge on land surface on term of influence to E.M.

PART II: EXTERNALITY OR ASTRONOMY ON E.M

On the outward wise from land surface, the first we should consider is atmosphere which we have done in previous chapters. The second is Van Allen belt which is established approximately from 400 miles to about 36 000 miles from Earth's surface. With plasma sphere attached to, the Van Allen belt makes itself be a very interesting theme to research. The next is the Moon, unlike any other planet or star in the universe, the Moon is rather close to Earth and to exert magnetic influence on it.

Unlike sea & land which are on the Earth, moving with the Earth; the astronomic bodies are not on Earth and especially assumed as on no-move or stationary during our problem solved. After this part, we will realize who is who among those who influence the Earth.

This part is a long discussion about several huge issues; we again recall the Biot-Savart expressions 9.1.19 and 9.1.20:

Quote:

$$d\vec{B} = \frac{\mu_0}{4\pi} \frac{(nAq|\vec{v}|)d\vec{s} \times \hat{r}}{r^2} = \frac{\mu_0}{4\pi} \frac{(n\,A\,ds)q\vec{v} \times \hat{r}}{r^2} = \frac{\mu_0}{4\pi} \frac{(dN)q\vec{v} \times \hat{r}}{r^2} \quad (9.1.19)$$

where r is the distance between the charge and the field point P at which the field is being measured, the unit vector $\hat{r} = \vec{r}/r$ points *from* the source of the field (the charge) *to* P. The differential length vector $d\vec{s}$ is defined to be parallel to $\vec{v}$. In case of a single charge, $dN = 1$, the above equation becomes

$$\vec{B} = \frac{\mu_0}{4\pi} \frac{q\vec{v} \times \hat{r}}{r^2} \quad (9.1.20)$$

Unquote

Under the light of Biot-Savart, we realize that the Van Allen belts and Moon can exert their electric forces and certainly induce magnetism on the Earth. We are going to discuss and debate the basic questions about them. The discussion is conducted in the forms of "problem set & solve" and prolonged in chapters.

CHAPTER I: VAN ALLEN BELTS

I-General about Van Allen belts:

I and many among my colleagues used to view Earth's magnetic field as one that can only facilitate the magnetic compass to indicate the North magnetic direction and no more it can offer. But the Van Allen belt appears to change our mind; the Earth's magnetic field is not very strong but good enough in facilitating the establishment of Van Allen belt which can stop any fast moving charged particle. The Van Allen belt, as matter of fact, plays an important role in protecting our planet.

(The following is almost prepared with close reference to SPACE & NASA's documents and papers, which are open to everyone and nothing is from classified information.)

1-What is Van Allen belt: a collection of charged particles, gathered by Earth's magnetism, in a shape of belts around Earth.

2-The earliest discovery to the belts: 1958 by Explorer 1 (NASA's Satellite).

-Belt's structure: normally 2 belts, one is separated from the other by a large drain/gap; can split up to 3 belts or merged in one sometimes. The inner belt is almost close to Earth surface at minimum about 400 miles; while the outer belt is rather far and even stretches to 36000 miles.

The drain is discovered in 2012; a remarkable feature of the belt is that its drain can stop the ultra-fast electron in there. The scientists from NASA are still not determined about reason for the belt separation. Their hesitation is quite understandable because no mechanism of electric force or magnetic field is found linking to the wax & wane activity of the belts (which is in conjunction with Sun's activity). The scientists likely ascribe to the particles from space or the Sun.

The unidentified particles or quarks might be gathered among the belts and create an unknown field between them, the field makes the belts split-up wide.

- How the belts work: They can wax and wane in response to incoming energy from the Sun, sometime swelling up enough to expose satellites in low-Earth orbit to damaging radiation.

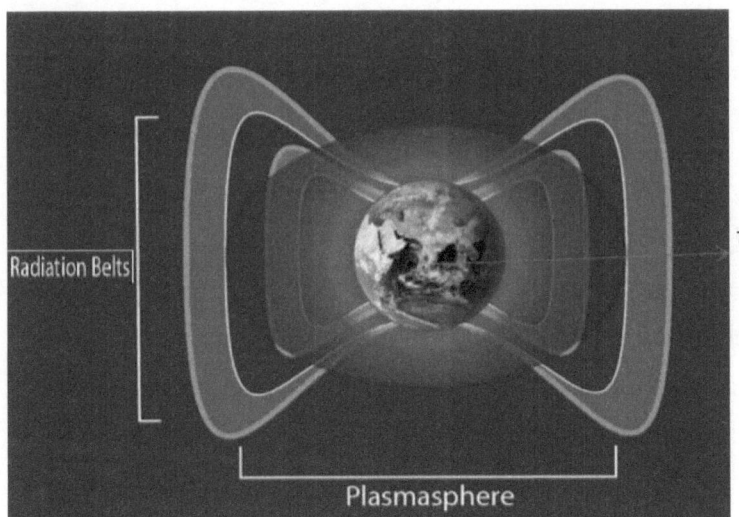

Figure 1/III-Van Allen belts-(*Credited to* NASA)

- Undiscovered about the belts: With some satellites working on space at all the time; the more research to the belts are carried out, the more expect to be discovered; even the unidentified particle or quark.

- Plasmasphere: A companion to the belts is a giant plasma sphere which can be assumed as twinned to the belts. The plasma sphere is cloud of relative cool and charged particles that fills the outermost region of Earth's atmosphere. That sphere begins approximately at 400-600 miles (rather close to Earth surface) and extends partially to outer Van Allen belt.

In scope of this book, the belts and plasma sphere make a thick charged layer around the Earth; the thickest is on the plane perpendicular to Earth rotation axis and expanding through the Sun. The thick layer can play a role of a charged ring whereby the Earth is rotating in; on the other hand, the layer is a bar that restricts the Moon's electric influence to the Earth.

II-Van Allen belts on E.M:

1-Charge in the belts: We are noticed that the fast-moving electron is stopped and swallowed in the belts or their drain; as matter of fact, such behaviour indicates that there is more likely positive charge in there. However, the electron condensation there can reach a certain level by which the belts become neutral then turns to negative-the belt charge reversal.

2-How do the belts influence on E.M: The Earth is rotating inside the belts and the plasma sphere; the problem becomes one of "globe rotating inside another globe" and certainly it must be a hard problem. Definitely a certain magnetic amount must be induced by the belts and plasma sphere.

Nonetheless, there is no data available or reliable for this problem. On the other hand, even if the data is found; so we set and solve a problem that totally upon assumption. Van Allen belt and Plasma sphere are rather complicated; we may divide them into many hoops approximately. Set and solve problem for each hoop and sum up the results.

PROBLEM

If total charge of the Van Allen belts and plasma sphere is equivalent to a single band of $+10^6$ Coulombs around Earth equator at a distance 1500 km from Earth surface, and evenly distributed on it.

How is the magnetism induced by the Band to the equator (Earth)?

1/The proposed data:

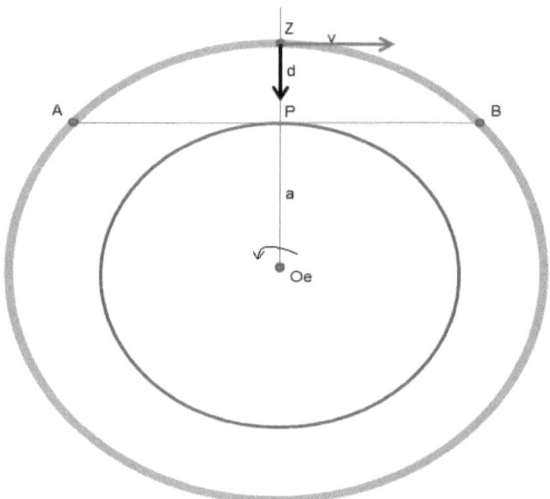

Figure 2/III-Van Allen belt to the Earth (not drawn in scale)

- The equator is presented by a blue circle as a hoop around centre Oe, average radius is OeP=a=6356.8 km.

- The yellow circle around the same centre Oe is a band that displays the equivalent circle or the Band for Van Allen belts and plasma sphere incorporated.

- OeP is extending beyond P and intersecting with yellow circle at Z (d=1500 km distance).

- AB is horizontal line for P.

- The Earth rotation (from West to East) fabricates a relative rotation of the Band around the Earth from East to West like Sun's wise.

2/Assumptions:

-Every move is viewed as standing from North side of the Earth.

-The Band is charged with a given value $Q=+10^6$ Coulombs, pure positively, continuously and evenly along the Band.

3-__Solution__: (Ref. to the figure 5/II)

-Every point from the charged band can induce magnetism to "P" at all the time. Nonetheless, the magnetic intensity induced by each charged point definitely reaches max when the distance from "P" to the charged point is shortened to minimum or inversely proportional to distance (P to Z). *(P to Z is the shortest distance from P to any point on the Band, isn't it? This is a problem of geometry and certainly not included in this discussion)*.

Thus, the Biot-Savart expression 9-1-20 is applied:

$$\vec{B} = \frac{\mu_0}{4\pi} \frac{q\vec{v} \times \hat{r}}{r^2} \qquad (9.1.20)$$

Where $\vec{v} * \hat{r}$ is the product of 2 vectors, the speed "v" and unit vector "r"; its direction is to come outward straight to reader, or true North for this problem.

One million Coulombs are distributed evenly and consecutively along the Band (as assumption).

The charged points, one after another, are continuously passing "Z" whereby it induces the max magnetism at P. The continuous rotation is to maintain the max of induced magnetism at "P" at all the time.

The calculation can be as following:

- For non-directional quantity of magnetism with any charged point passing Z:

$$/\vec{v} * \hat{r}/ = v*\sin(90°) = v;$$

and the "r" in this circumstance is $r = PZ = 1500$ km $= 1500*10^3$ m

+ Speed "V" of yellow band or every charged point:

$V(z) = 2\pi*(6356.8+d)*10^3/24*60*60$ m/s

$V(p) = 2\pi*6356.8*10^3/24*60*60$ m/s

$V = V(z) - V(p) = 20\pi*d/24*36 = \pi d/43.2$ m/s

(= $2\pi*1500*10^3/24*60*60 = 109.083078$ m/s if we need to produce it)

+ Charges distributed on the Band for every 01 second of time:

$q = Q/24*60*60$

(= $10^6/24*60*60 = 11.574074$ Coulombs if we need to produce it).

- The permeability of the media between Z and P, it is supposed to be one for normal air, the μ is approximately:

$\mu_0 = 4\pi*10^{-7}$

Thus:

$\Delta(3) = (4\pi*10^{-7}/4\pi)*(Q/24*60*60)*(2\pi*d*10^3/24*60*60)*(1/10^6*d^2) =$

$= 2\pi*Q/24^2*36^2*10^8*d = 8.4169*10^{-23}*\dfrac{Q}{d}$

$$\Delta_{(3)} = 8.4169*10^{-23}*\dfrac{Q}{d} \quad \text{(V-A 01)}$$

Where unit of "Q" is in Coulomb and unit of "d" is km.

Replace Q=10^6 Coulombs, d=1500 km into the above

formula:

$$\Delta(3) = 8.4169 \times 10^{-23} \times 666.666667 = 10^{-20} \times 5.611267 \text{ Tesla}$$

(Note: *the Belt is assumed as on a plane perpendicular to Earth's rotating axis, the assumption is not verified and the reality definitely is not be so. Therefore, in order to calculate for a better accuracy, the belt as well as the plasma sphere can be represented by many bands (or hoops) in a larger problem*).

Conclusion:

If Van Allen belt and plasma sphere are equivalent to a band of Q Coulombs (the Band) around the Earth on plane of Earth's equator. The Band is to induce magnetism on Earth's equator; its additional intensity is calculated as (V-A 01):

$$\Delta_{(3)} = 8.4169 \times 10^{-23} \times \frac{Q}{d}$$

In reality, Van Allen belts and plasma sphere are not only changing in shape and size, but also in intensity. With such a complicated change, the above mentioned "Δ3" demonstrates nothing but "yes" to the question "influence?". Furthermore, as noted above, the larger problem is required to estimate the additional magnetic intensity with better accuracy.

CHAPTER II: TWO OPPOSITE HEMI-SPHERES OF MOON

Moon is the only satellite to Earth. In human life, the Moon merges in poet, music and children's daily dream.

Since 50th of last century, human being had succeeded in researching the Moon, U.S scientists even launched in there and planned to build an outpost on it, their plan is still going on. Nonetheless we don't discuss about their plan, but we do discuss about another issue: **the charges on Moon and their influence to the Earth.**

We denote magnetic compositions contributed by Moon's hemisphere the $\Delta 4$, $\Delta 5$ and $\Delta 6$ where $\Delta 4$ is for negative hemi-sphere influence, $\Delta 5$ is for positive hemi sphere; and $\Delta 6$ is for total charge contribution.

The following is charged Moon that depicted by NASA:

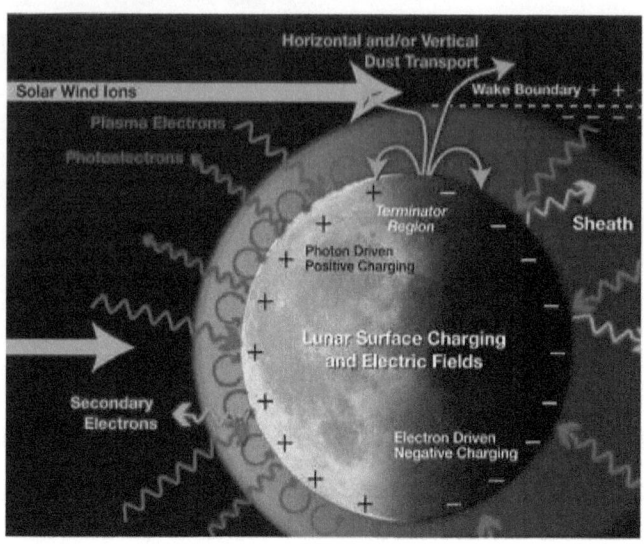

Image 1/IV-Moon image (Credited to NASA)

The ion wind from Sun is striking on Moon's surface and energizing the electrons on there, as NASA's argument, the direct result is to electrify the shined part positively. The highly energized electrons flock to settle on the dark side of the Moon where they lost their energy. That's how the Moon surface is such illustrated in the Image 1/v. In order to clarify the Moon's hemisphere effects on Earth, we set and solve the following problem.

1-<u>Problem 1</u>: The lunar hemisphere effect.

Under sunlight, the Moon is a charged or an electrified object. Suppose that the nature of object is neutral and so the positive charge on shined side is balanced against the negative charge on the dark side.

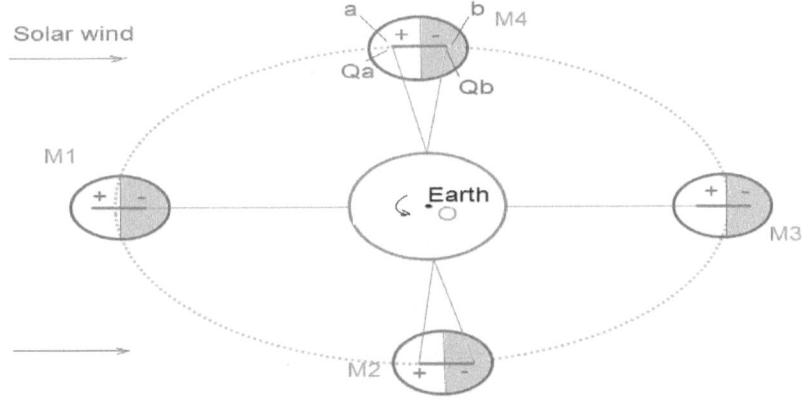

Figure 2/IV-Solar Wind and Moon charge

With suggestion from NASA's image; consider the magnetism induced by Moon under the light of Biot-Savart laws.

1/1-<u>To understand more about the problem (general analysis)</u>:

The Moon's influence on Earth is not general or with a common effect for everywhere but is to be determined to each specific position. The relative position of Moon in Earth coordinates is changing at all the time; and especially the Moon's position to each specific position of observer on the Earth is changing too. For this problem, Houston (30N, 95W) is selected as observer's position.

What is the Moon's magnetic influence on Earth, (restricted time in 1 day of Moon (29.5*24 hrs. for anything relevant to light and electric) within May, 2017)?

Data from Astronomy Almanac (or Moon timetable) for Houston (30N, 95W):

- Quarter 1, semi-Moon 49.1%, 02nd May 2017. Moon set/rise: 12.38 / 02.20 hrs. Distance to the Moon: 377652 km. Altitude at meridian passing: 75.7^0 at 19h31".

- Full Moon (99.4%) on 10th May 2017. Moon rise/set: 20.01/07.14 hrs. The Moon's meridian passing 00.54", distance: 389519 km. Altitude at meridian passing: 48.6^0).

- Quarter 4 (semi Moon (55.4%), 18th May 2017. Moon rise/set: 01.34/12.51 hrs. Distance to the Moon: 391827 km. Altitude at meridian passing: 62.6^0 at 05h56".

- New Moon on 25th May 2017. Moon rise/set: 06.27/20.14 hrs. Distance to Moon: 357292 km. Altitude at meridian passing: 76.0^0 at 13h18".

- Earth's average radius: 6356.8 km.

- Moon's average radius: 1738.1 km.

1/2-Assumptions:

1/2/a-The Moon is neutral but charged oppositely on 2 hemispheres, one is opposite, while the other is negative.

1/2/b-The charge is distributed equally on both two hemispheres.

1/2/c-The charge is distributed evenly everywhere on

surface of each hemisphere, and only on surface.

1/2/d-Each hemisphere is charged up to a certain value Qa or Qb; with |Qa|=|Qb| and both of them stay unchanged.

1/2/e-Within every 24-hr interval, the Moon is assumed to be stationary.

1/2/f-The distance (from Astronomy Almanac) from a designated observer to the Moon is one to the nearest point of Moon's surface.

1/2/g-The magnetism induced by Moon is stored on the Earth's surface and underneath, and so contributes to Earth magnetism.

1/3-Solution:

With assumption that the electric charge is distributed on Moon surface only, we consider the Moon as two hollow hemispheres; both of them are charged evenly either positively or negatively. The rule for gravity centre of a hollow hemisphere is applied to determine the charge centres. The mentioned rule states that the gravity centre of a hollow hemisphere is $4R/3\pi$ from the centre of designated globe/sphere. The following is an illustration to the law:

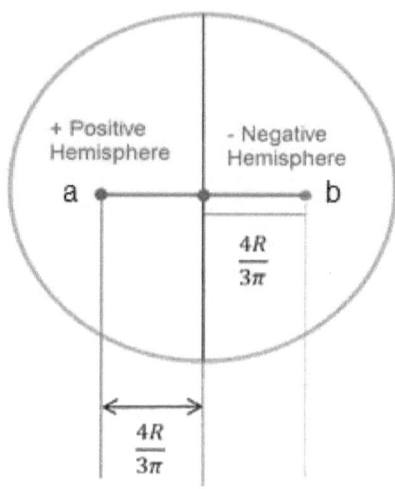

Figure 3/IV-Two opposite hemispheres of charges, (*Not drawn in scale*).

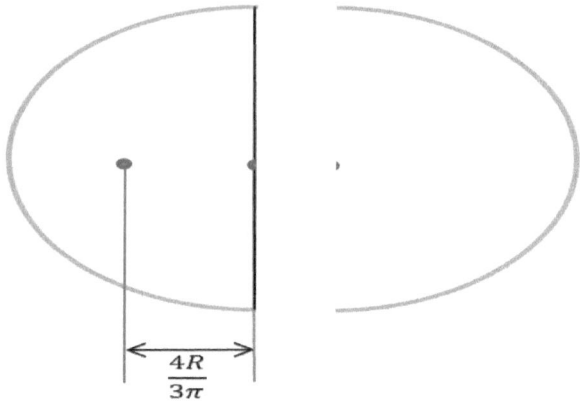

Figure 4/IV-Two hemispheres of Moon Figure, (*separated, not drawn in scale*).

The charge on Moon's hemisphere is (+Qa) and (–Qb),

each is designated for a respective centre of one hollow hemisphere. Distance from (a) to (b) is 2*4R/3π where R is average radius of Moon.

The following is an image of Moon and Earth on sky:

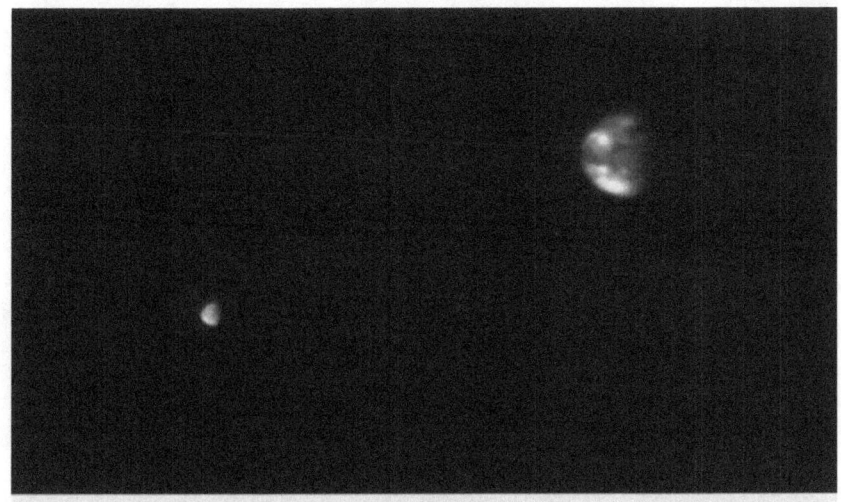

A view of Earth and its Moon, as seen from Mars. It combines two images acquired on Nov. 20, 2016, by the HiRISE camera on NASA's Mars Reconnaissance Orbiter, with brightness adjusted separately for Earth and the moon to show details on both bodies. Credit: NASA/JPL-Caltech/Univ. of Arizona.

Image 5/IV--Moon-Earth on Nov.20.2016 or about semi-moon.

HiRISE camera is near our Earth, the Moon in the image is rather far. Both Earth and Moon are shined on the hemispheres that face the Sun. The noted date is 20 Nov. 2016 when the Moon appears as semi-moon to Earth.

The following is a figure illustrating 8 phases of Moon around Earth in a day of Moon (29.5*24 hrs.).

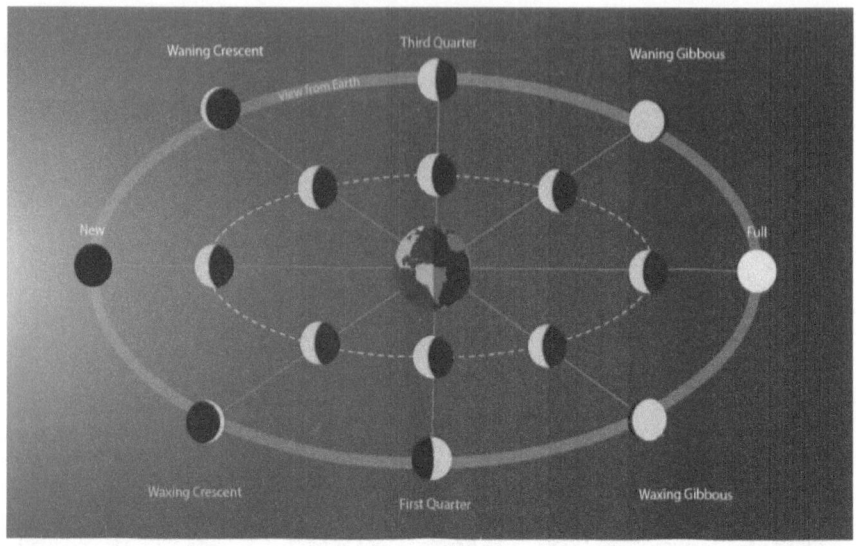

Figure 6/IV- Eight phases of Moon around the Earth in a day of Moon (29.5*24 hrs. and not 27.3*24 hrs.)

We will consider 4 from them: NEW and FULL where the Earth centre (O) and (a), (b) are aligned on a plane and assumed as one straight line. The semi-moon days on F.Q and T.Q when both (a) and (b) are at the same distance to Earth centre.

1/3/1-Solution 1: seek B(hs1).

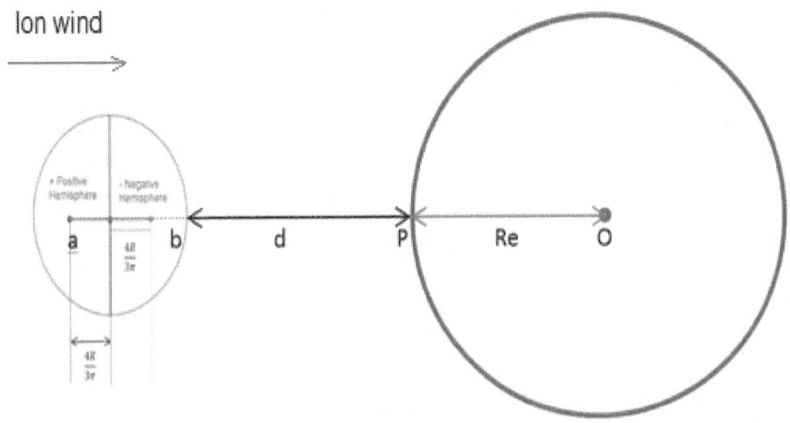

Figure 7/IV- Moon & Earth on new Moon (*Not drawn in scale*)

B(hs1) = B(hs.a)+B(hs.b),

Recall Biot-Savart 9.1.20, we have:

$$\vec{B} = \frac{\mu_0}{4\pi} \frac{q\vec{v} \times \hat{r}}{r^2} \qquad (9.1.20)$$

For the Moon hollow hemi-spheres, we are considering each of them as a hollow one because we did assume so without any potential dispute. Therefore, by 9.1.20, a certain charge numbered "i" is to contribute to Earth magnetism as following:

$$B(hs.i) = \frac{\mu(i)}{4\pi} * q_i(v_i * \sin(\alpha))/r_i^2$$

μ or μ_i is considered as μ_0 because we assume that the distance from Earth to Moon is rather far that dominates the distance from one to another centre of those two, Earth and Moon.

$$\mu_0 = 4\pi * 10^{-7} \text{ Therefore } \mu_0/4\pi = 10^{-7}$$

And we have magnetism contribution:

$B(hs.a) = 10^{-7} * Q_a * (v_a * \sin(\alpha))/r_a^2$

The angle between vector $\vec{v}$ and unit vector $\hat{r}$ is almost $90°$ approximately, therefore $\sin(\alpha) \approx 1$, so we have:
$B(hs.a) = 10^{-7} * Q_a * v/r_a^2$

We may replace the values of (d), R and (Re) into the following formula to calculate (r_a) as distance from (a) to Earth centre (O):

$r_a = Re + d + R(1 + 4R/3\pi) = Re + d + R(1 + 0.4244)$

- Another (r) for nearer hemi-sphere is:

$r_b = Re + d + R(1 - 4R/3\pi) = Re + d + R(1 - 0.4244)$.

The speed (v) is a direct proportional problem of speed of P and M against the distance from O to P, from O to M and from O to either (a) or (b) on M:

$v_p/v_m = PO/MO$ or $v_p/v_a = PO/aO$; or $v_p/v_b = PO/bO$;

With Earth average radius $PO = R_e \approx 6356.8$ km, P to Moon distance: $d = 377652$ km. Moon's radius: 1738.1 km, we have:

$aO = r_a = PO + Pa = 6356.8 + 377652 + 1738.1(1+0.4244) = 386484.54964*10^3$ m ;

$bO = r_b = PO + Pb = 6356.8 + 377652 + 1738.1(1-0.4244) = 385009.25036*10^3$ m;

$v_p = 410.611$ m/s (note that different latitude, different speed while the Earth's rotation is almost unchanged).

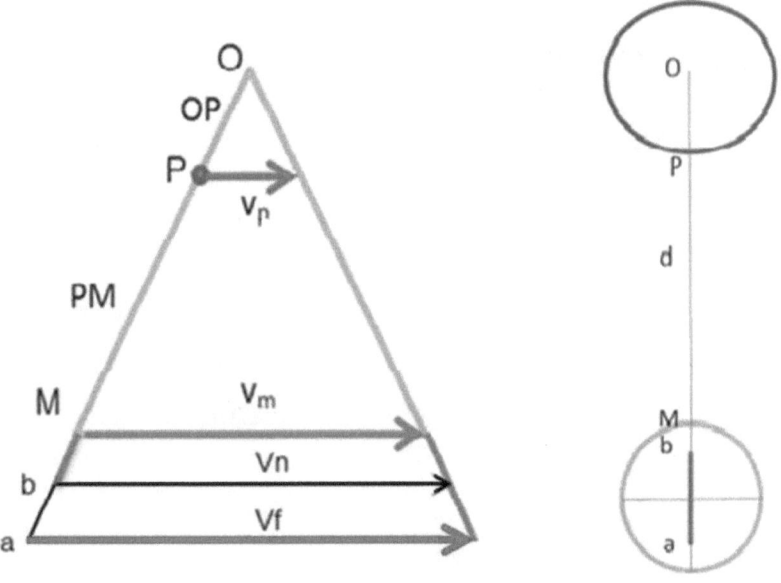

Figure 8/IV- Direct proportional ratio of "v" (*Not drawn in*

scale)

$V_a = 400.346*(PO+Pa)/PO = 400.346*386484.54964/6356.8$
$= 24340.4769$ m/s

$V_b = 400.346*(PO+Pb)/PO = 400.346*385009.25036/6356.8 = 24247.5637$ m/s

With $V_p = 400.346$ m/s and some other known data, the speeds of (a) and (b) relatively to P is:

$v_{ap} = 24340.4769$ m/s

$v_{bp} = 24247.5637$ m/s

The calculation for magnetism induced by Qa and Qb to Earth is carried out as following:

$B(bs.1) = B(bs.a) + B(bs.b)$

$B(bs.i) = 10^{-7} * Q_i * V_i * \sin(a) / r_i^2$

We do assume that $\sin(\alpha) \approx 1$, and (Qi) represents either Qa or Qb, respectively we have:

$B(bs.b) = 10^{-7} * 24247.5637 * Qb / r_b^2 =$

$= 10^{-7} * 24247.5637 * Qb / 385009.25036^2 = 1.63578 * 10^{-20} * Qb$

$B(bs.a) = 24340.4769 * 10^{-7} * Qa / r_a^2 =$

$= 24340.4769*10^{-7}*Q_a / 386484.54964^2 *10^6$

$= 1.629539*10^{-20}*Q_a$

We replace the above B(hs.a) and B(hs.b) into B(hs):

$B(hs) = 1.63578*10^{-20}*Q_b + 1.629539*10^{-20}*Q_a$

We recall the basic B(hs) for another negotiation about vector B:

$$\vec{B} = \frac{\mu}{4\pi} *q*\vec{v}*\hat{r}/r^2$$

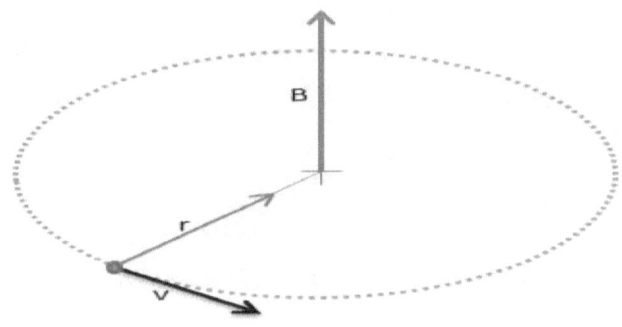

Figure 9/IV-Direction of vector B

Although "Q" is not directional quantity but its value is either (+) or (-), then "Q" value can name the direction of vector B. The B(hs) in this problem is one of prominent examples of the role of (Q) in vector B direction. As matter of ion wind's effect, Qb is negative or (-) while Qa is positive or (+). Thus, $B(hs.b)$ is negative while $B(hs.a)$ is positive, the total $B(hs.a)+B(hs.b)$ is calculated as following:

Replace $Q_a = -Q_b$ in to the above formula of $B(hs)$

$B(hs) = 1.63578*10^{-20}*Q_b - 1.629539*10^{-20}*Q_b =$

$= 6.241*10^{-23} Q_b$ Tesla

Conclusion for solution 1:

The opposite hemi-spheres of Moon induce magnetism on P(Houston). As seen in the figure 6/v above, Qb is nearer to Earth; by the inverse law of electric force, Qb can influence more than Qa.

Moon is considered as stationary during the Earth is rotating, so Moon's magnetic vector is parallel with Earth's rotation axis and consists of X and D compositions.

Qb might be negative which induces a contribution that heads to North, against current Earth's magnetism.

The result of calculation ($6.241*10^{-23}*Q_b$ Tesla) is not too tiny because Qb is still unknown and varying at all the time. If we give Qb or Qa a certain figure such as 10 6 Coulombs, we may have B(hs), but we don't do it now as our discussion is still going on.

1/3/2-Solution 2(p3-full Moon):

We are to solve the problem for the same location of observer as 1/3/1 but for full Moon (phase 3 or p3) with the following entries:

-Full Moon on 10th May 2017, the Moon is passing meridian at 00.54'; distance: 403932 km. Altitude at meridian passing: 48.60°).

B(hs.3)=B(hs3.a)+B(hs3.b),

Recall Biot-Savart 9.1.20, we have:

$$\vec{B} = \frac{\mu_0}{4\pi} \frac{q\vec{v} \times \hat{r}}{r^2} \qquad (9.1.20)$$

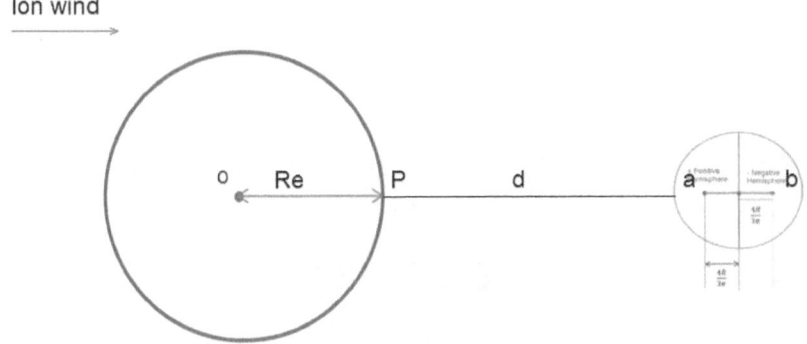

Figure 10/IV-Full Moon effect.

- Bring some from last problem forward:

> Earth average radius Re=6256.8 km.
>
> Moon's average radius R=1738.1 km.
>
> Distance from either (a) or (b) to Moon's centre 0.4244 km;
>
> The "d" on full Moon for this calculation is 403932 km.

We may replace the values of (d), (Re) and (R) into this formula to find value of (r_a) as distance from (a) to Earth centre (O):

$r_a = Re+d+R(1-4R/3\pi) = Re + d + R(1-0.4244) =$
$6356.8+403932+1738.1*0.5756 =$

$= 411289.25036$ km $= \underline{411289.25036*10^3 m}$.

- Another is (r_b) for further hemi-sphere is:

$r_b = Re+d+R(1+4R/3\pi) = Re+d+R(1+0.4244) =$
$6356.8+403932+1738.1*1.4244=$

$= 412764.54964$ km $= \underline{412764.54964*10^3 m}$

The speed (v) is a direct proportional problem of speed of (P) and (a) or (b) with the distance from (O) to (P), from (O) to (a) and from (O) to (b) on (M):

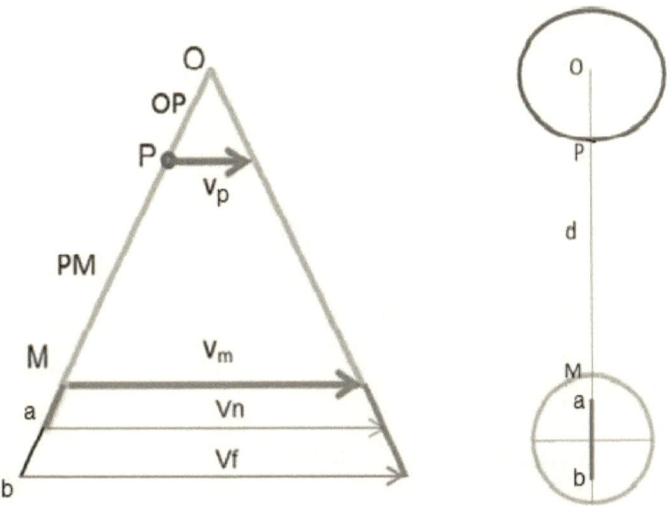

Figure 11/IV-Direct proportional between velocity (v) and distance to (O).

(Not drawn in scale)

$v_p/v_a = PO/aO$; or $v_p/v_b = PO/bO$;

With Earth average radius

$PO = R_e \approx 6356.8$ km;

Distance from (P) to Moon: $d = 403932$ km;

Moon's radius: 1738.1 km;

$aO = r_a = PO + Pa = (6356.8 + 403932 + 1738.1 \ast 0.5756) \ast 10^3$
= $\underline{411289.25036 \ast 10^3 m}$;

$bO = r_b = PO + Pb = (6356.8 + 403932 + 1738.1*1.4244)*10^3$
$= \underline{412764.54964*10^3 m}$;

$v_p = 400.3458$ m/s,

$Va = 400.3458*Oa/PO =$

$= 400.3458*411289.25036/6356.8 =$

$= \underline{25902.643463}$ m/s

$Vb = 400.3458*Ob/PO =$

$= 400.3458*412764.54964/6356.8 =$

$= \underline{25995.556544}$ m/s

$B(hs.3) = B(hs3.a) + B(hs3.b)$

Assume that $\sin(\alpha) \approx 1$, with (Qi) represents either Qa or Qb, respectively we have:

$B(Qi) = 10^{-7}*Qi*Vi*\sin(\alpha)/r_i^2$

$B(hs3.a) = 25902.643463*10^{-7}*Qa/r_a^2$

$= 10^{-7}*25902.643463*Qa/411289.25036^2*10^6 =$

$= 1.531262*10^{-20}*Qa$

$B(hs3.b) = 25995.556544*10^{-7}*Qb/r_b^2 =$

$$= 25995.556544*10^{-7}*Qb/412764.54964^2*10^6=$$

$$=1.525788*10^{-20}*Qb$$

$B(hs.3) = B(hs.a)+B(hs.b) =1.531262*10^{-20}*Qa+1.525788*10^{-20}*Qb$

We replace $Qb = -Qa$ to the above:

$B(hs.3) = B(hs.a)+B(hs.b)=1.531262*10^{-20}*Qa-1.525788*10^{-20}*Qa=$

$$=5.474*10^{-23}*Qa \text{ Tesla}$$

Conclusion for solution 1/3/2:

-When full Moon is viewed from Earth; the shined hemisphere is positive and nearer to Earth, its influence to Earth is more than the dark one. The additive magnetic vector at observer (P) is parallel with Earth's rotation axis.

The nearer hemisphere is inducing a contribution to Earth magnetism, which heads to South. Such heading is supportive to the current Earth magnetism.

1/3/3-Solution 3: (Semi Moon, B(hs.2) and B(hs.4))

The following p2 and p4 are semi-moon and viewed as neutral positions; where Oab is isosceles triangle, and so one of the two effects of the hemispheres has the same but opposite value with the other and eliminates it.

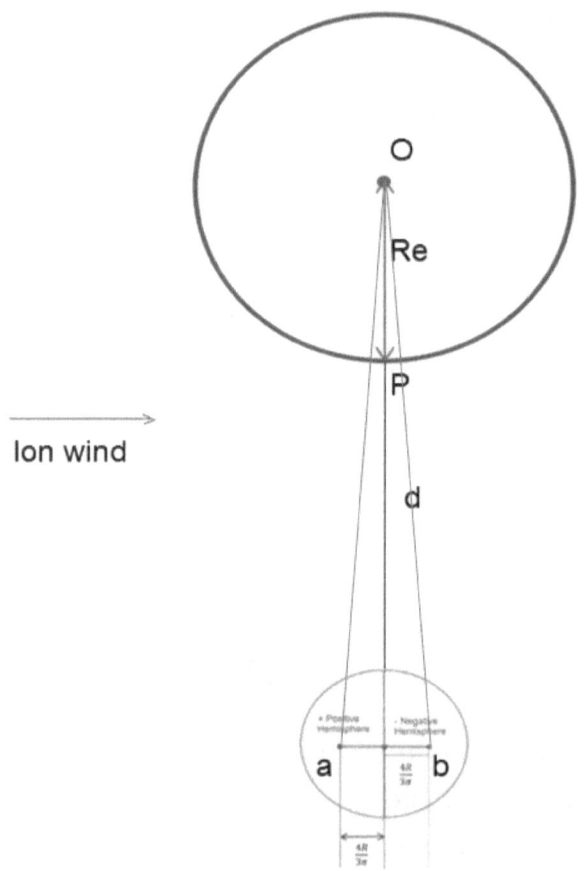

Figure 12/IV-Quarter 1 at p2, semi-moon status.

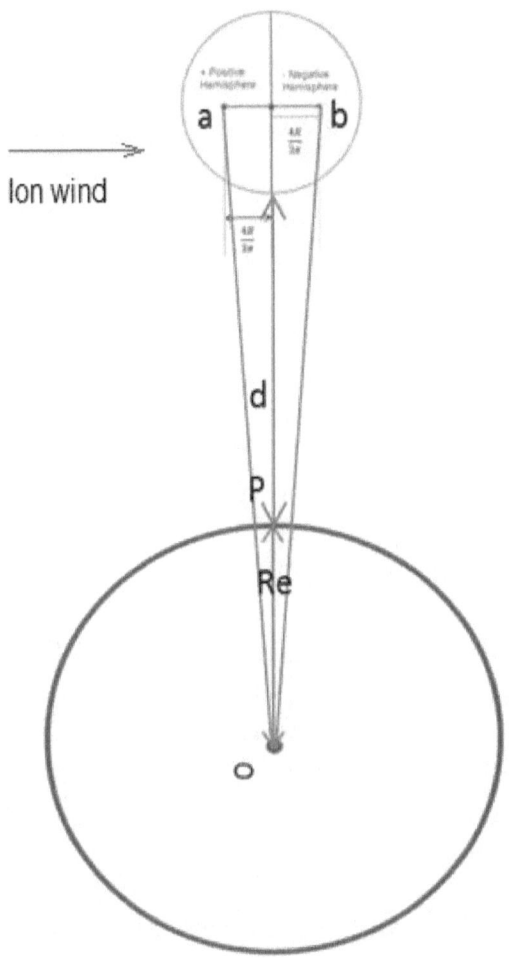

Figure 13/IV- Quarter 3 at p4, semi-moon status.

Nonetheless, that is not everything about those 2 positions.

After p2, the Moon keeps moving to p3; from p2 to p3, the shined part of Moon is viewed larger and larger until full Moon.

From p4, the Moon keeps moving back to p1 (new Moon), meantime the shined part of Moon becomes smaller and smaller until new Moon. We got to review the 8-phase day of Moon:

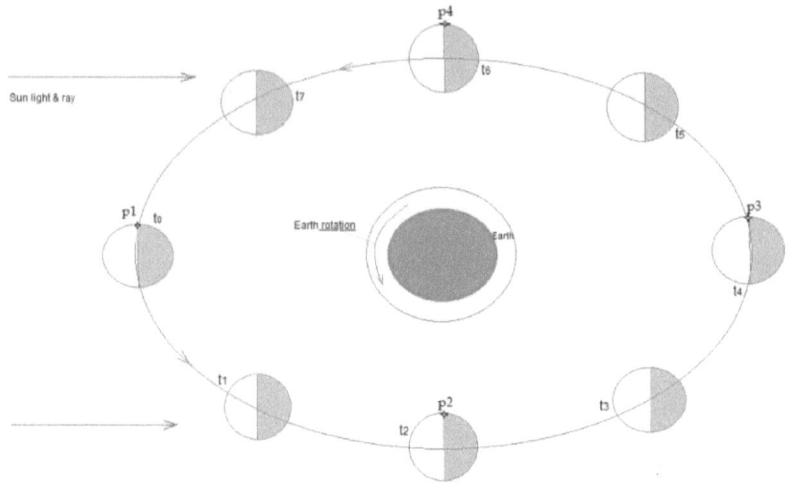

Figure 14/IV - Important phases of Moon

Conclusive words for 4 solutions about Moon:

With a quick review at all of 4 contributions from Moon as above, we can realize that every question of this problem can be solved rapidly with Biot-Savart Expression 9.1.20. For the time being, the bilateral hemisphere Moon's effect to Earth's magnetism is in brief as following:

The Moon at new-moon position induces magnetic effect on any position on Earth surface where Moon's electric field may exert on. The new-moon's effect is against the Earth magnetism while it is supportive on full-moon position when the Moon's shined hemisphere is nearer to Earth than the dark one.

At both semi-moon positions, the Moon's hemisphere effect is quite neutral or nil.

And that:

From 2nd semi-moon (p4) up to the 1st semi-moon (p2) through new Moon (p1), the Moon's hemisphere effect on any observer on Earth surface is against Earth's magnetism, but it is reverse from 1st semi-moon (p2) to 2nd semi-moon (p4) through full Moon (p3), the Moon's hemisphere effect is supportive to Earth magnetism.

CHAPTER III: MOON'S TOTAL CHARGE EXERTING ON EARTH

Every previous solution is solved with many assumptions. Anyhow, the results are meaningful; they do confirm that the semi-moons in every day of Moon are approximately the reverting points when the Moon's hemisphere effect changes from negative to positive or vice versa. Hence the two semi-moons divide the Moon's orbit into 2 different, opposite halves; the effect on one half is going against or contrarian the current E.M, while the other's effect is supporting or backing the current E.M.

The radiation from the Sun is changing and the Moon charge therefore is changing accordingly. So there must be a sequent change in Moon's effect on E.M, and this chapter is a discussion about the change in circumstance when Moon's charge is changing and the total charge "Qa+Qb" is no more a "0".

We leave the influence of permeability of the media between Moon and Earth centre for another problem; just consider the Moon's influence on Earth's surface.

Let's set and solve the following problems for Moon's effects.

1-Problem of coaxial circles:

The Earth's rotation is to fabricate a relative rotating move on a wise opposite to the rotation of Earth such as the move of Sun or Moon. Therefore, we need to go through this argument before the any other. My way is to set and solve a problem.

a/ <u>Problem</u>:

A couple of circles (or two coaxial circles) with a common centre "O", a point "P" that is fixed or stationary on the inner ring of radius "a", while an object "C" is orbiting "O" on the outer ring, at "d" far from inner ring.

-What identifies those circles?

-Relation between the given figures?

b/ <u>Solutions</u>:

In order to have some further understanding about the figures given, such as radius "a", distance "d", we need to analyse them in a mock figure (Figure 1/IV).

As matter of facts, the distance "d" and the ratio of its own value on the radius "a" of inner ring is definitely to identify the status of the general image whereby the figures built.

How do they work?

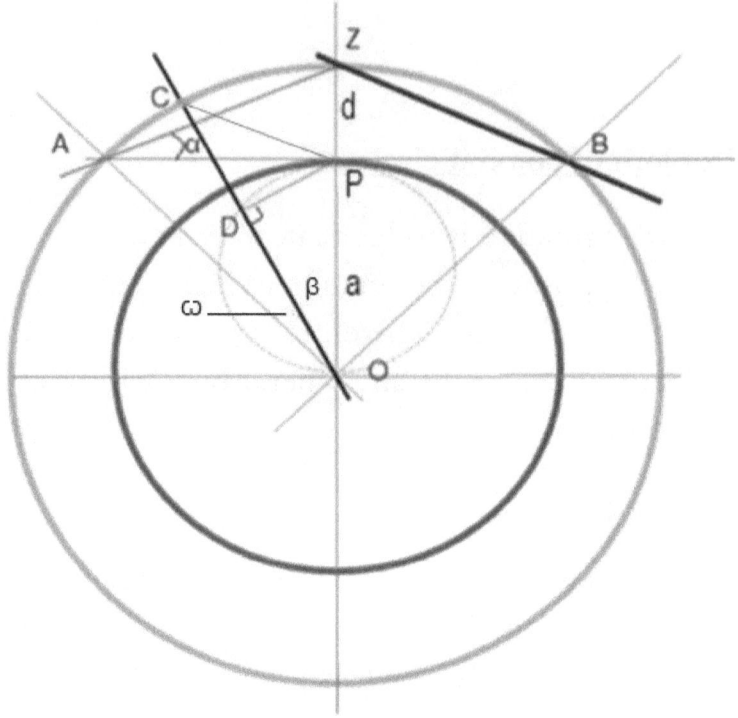

Figure 1/V-Mock figure of given data

Have a reference to the above mock figure, we find out:

OA=OZ=OC=OD+DC=a+d as radius of the outer circle.

- In the right triangle APO, the Pythagorean rules are applied:

AP=(OA²-a²)^{1/2} = ((a+d)²-a²)^{1/2}

- And in the right triangle APZ:

Tan(α)=d/AP=d/((a+d)²-a²))^{1/2}

Tan(α)=$\sqrt{d/(d+2a)}$ or

$$\sqrt{\frac{d}{d+2a}}$$

- The equal angles: AOZ=ZOB=2α

- Angle ⓐ: the angle between OA and OC, this is to vary when "C" moves along the outer ring.

- In right triangle PDO: PO=a, β+ⓐ=2α or β=2α-ⓐ; and PD=a*sin(β), OD=a*cos(β).

- In right triangle PDC: CD=OC-OD=(a+d)-(a*cos(β))

CP²=CD²+DP²

Angle CPD: tan(CPD)=(a+d-a*cos(β))/a*sin(β)

Conclusion:

- For every couple of coaxial circles with given radiuses, we always find "α" *(with the above formula of Tan(a))* and certainly the other angle of the right triangle is (90⁰-α). Although we don't apply angle *(a)* in this problem, but that angle is always the identity of any coaxial circles.

- There are always equal angles: COZ=APD, and AOB=4α

(Note: PD is drawn perpendicular to OC, then we realize that "D" always hugs OP with a right angle ODP. Therefore "D" is always on a circle of diameter OP while "C" is moving along the outer circle. This is a locus problem for entertaining the geometric lovers, and not included in this discussion.)

2-Problem of Moon's total charge exerting directly on

Earth surface:

The following figure 2/IV is for a pure theoretic problem. The Moon's charge is assumed: $q = 1*10^6$ Coulombs.

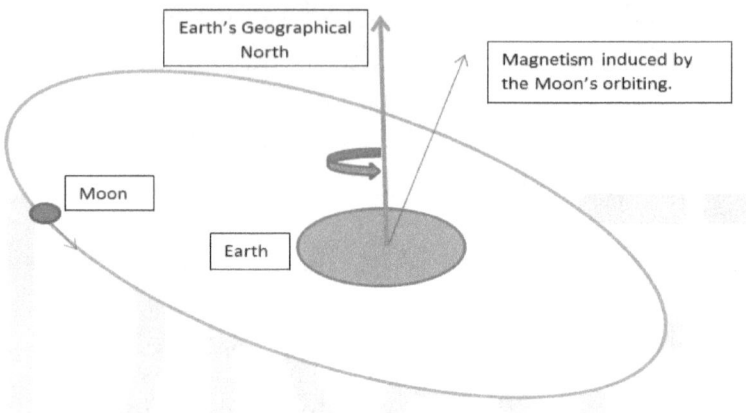

Figure 2/V-Moon's orbiting and its magnetism induced on Earth

Biot-Savart expression (9-1-20) is recalled and the magnetism contributed by Moon's orbiting is calculated right away with some Moon's facts as following:

$$\vec{B} = \frac{\mu}{4\pi} * \frac{q * \vec{v} * \hat{r}}{r^2}$$

With $\mu = \mu_0$, $q = 10^6$ C, $v = 1.022$ km/s, $r = 370400$ km we can obtain value of B. Nonetheless, our discussion is not about a pure theoretic problem, so we leave it and work with our problem for Houston (30N, 95W).

2/-The problem of Moon's total charge for Houston:

Houston of U.S chosen as observer in this problem is quite in the middle of Van Allen belt and plasma sphere; therefore it is very hard to detect the Moon's effect. Nonetheless, we keep working with that location as observer and the following is the problem:

Houston (30N, 95W), USA on 02nd May 2017, Moon's total charge Qm and other parameters:

- Illumination: 49.1%≈50%, this is to make sure that each of the two hemispheres is capable to eliminate the other, so the hemispheric effect can be null.

- Meridian passing: 19.31 hrs.

- Moon rise/set (excerpt from Moon's time table):

2	-	01:32 ↖ (290°)	12:38 ↗ (71°)	19:31 (75.7°)	377,652	49.1%
3	-	02:20 ↖ (287°)	13:38 ↗ (74°)	20:23 (72.6°)	382,856	60.2%

- Distance: 377652 km from point P to Moon.

- Moon's average radius: R=1738.1 km.

Question:

With Moon's total charge Qm≠0, consider the amplitude of magnetism that directly induced by Moon on Houston at 19.00 hrs.

2/1-To understand more about the question (analysis):

- The meridian passing is on 19.31 hrs at 75.7°.

- From the Moon's timetable: Moon rise/set is 12.38/(next day)02.20 hrs. or 12.38/26.20 (or Moon's on sky within 822 minutes).

- The distance from A or B to a general P (on longitude of P) can be viewed as the same and depicted in the following figure:

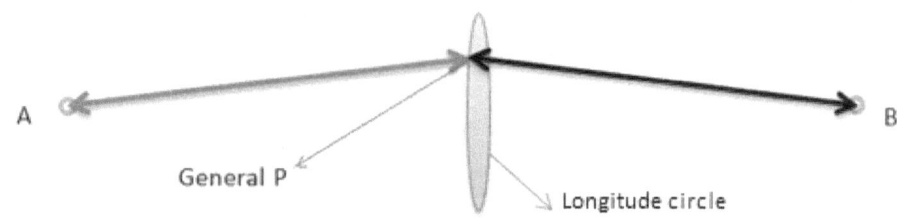

Figure 3/V-A general P and A-B

- The given P is on land surface with eye height "0", the horizon viewed from P is a plane with the straight line from Sunrise (A) to Sunset (B).

- Note that $AP = PB$, by Pythagorean theory $AP=(AO^2-OP^2)^{1/2}$ (Figure 4/IV), where $OP=Re$, the value of Earth average radius despite "O" is not coinciding on Oe (Figure 5/IV).

- A circle with centre (O) is designed to meet $AP=(AO^2-OP^2)^{1/2}$ and distance from P to Moon 377652 km. This centre (O) may not be coinciding on the terrestrial geometric centre (Oe).

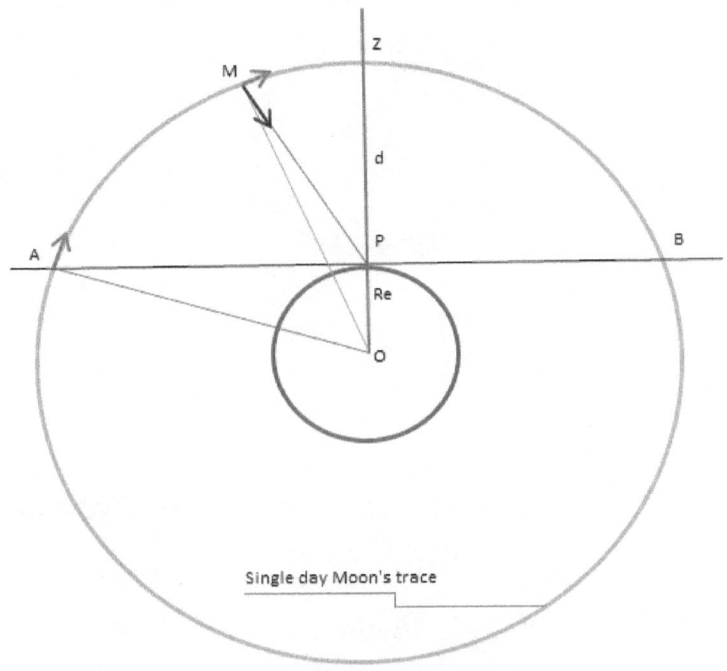

Figure 4/IV-Basic data of problem on plane of Moon's trace (*Not drawn in scale*).

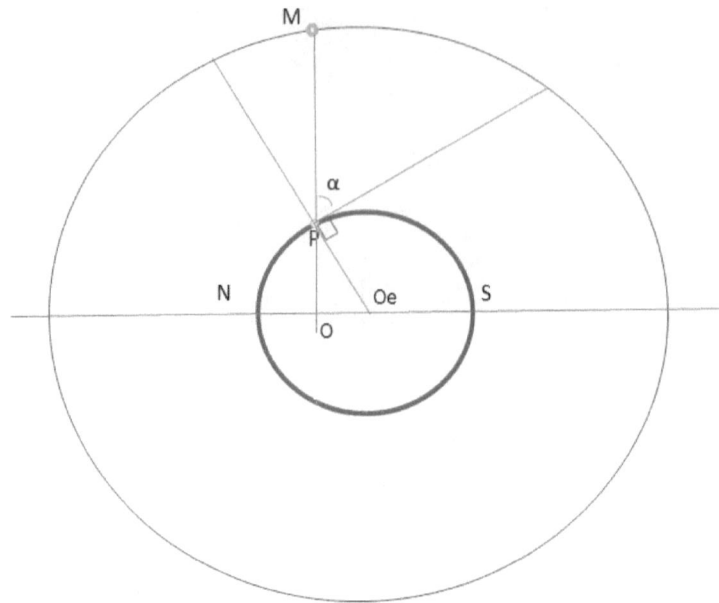

Figure 5/V—Moon at M. passing moment, Earth and P on meridian plane

(Note: the ring with M in Figure 5/IV is not 24-hr Moon's trace but meridian circle. The (O) is centre of the circle of Moon's trace in Figure 5/IV, while (Oe) is Earth's centre).

2/2-<u>Solution</u>:

2/2/1-<u>Step 1</u>: to re-think about the real things.

Like any other electric current or moving charge, the Moon definitely contributes magnetism. Moon's total charge has been assumed as neutral with 2 opposite hemispheres in previous problem.

But in reality the total charge of any object under solar wind can't be neutral at all the time; it is actually changing, and is given a value Qm (positive) in this problem.

The questions are how magnificent is the magnetism induced by the lunar total charge? Which wise (north or south) does it direct to?

2/2/2-Step 2: Assumptions.

- P (30N, 95W) is assumed as the stance for observer with eye height "0".

- Distance from P to A is the same as P to B or any position on the longitude of P to those 2 positions of Moon.

- The Moon is assumed as on "no real move" throughout the time of calculation. By this condition, the Moon's real move is not considered. Instead, we consider the Moon's relative move or 12(+) hr. trace around the Earth.

- Moon's total charge (Qm=Qa+Qb and Qm≠0) stays unchanged throughout this calculation.

- Distance: 377652 km is from P to the nearest position of the Moon. The distance from P to charge centre of Moon can be approximately added with ½ of Moon's average radius 1/2*1738.1 km.

The distance d=378521 km is assumed as distance from observer to Moon's charge centre.

3a-Step 3a: Approximate Solution (for student reference only).

Instead of bearing Biot-Savart from beginning, we can take cross path with an expression from the last problem for a charge moving on a straight line as following:

> The product of 2 vectors ($\vec{v} * \hat{r}$) is a vector that comes toward the reader, and its non-directional value is:
>
> $$|v|*|r|*\sin(\alpha) = v*1*\sin(\widehat{DCk}) = v*\cos(\alpha)$$
>
> Thus, the amplitude of magnetic field at D is:
>
> $$|B| = 10^{-7}*q*v*r_0^{-2}*\cos^3(\alpha) \text{ Tesla} \qquad \text{(F-m/v)}$$

The expression (F-m/v) is applicable for a charge moving on a straight line. If we assume such short 31-minute period as a short time, we may consider the Moon's move as one on a straight line. Let's replace q=Qm, and attribute the following quantities into that expression:

$r_0 = d = 378521*10^3$ m

$\alpha = 07.95°$ (this angle can be converted into $07° 57"$)

<u>To seek for</u> (v):

The Earth rotation around its own axis is to fabricate a relative move of Moon around the Earth as seen from everywhere in the world. The target of this paragraph is to find out the relative speed of Moon on its daily trace-from Moonrise till Moonset on the sky.

The relative speed of Moon depends on observer's position

on the Earth. For our problem, we can find the Moon's speed by interpolating from observer's distance to Moon and her/his speed around Earth rotation axis. The following is the calculation:

MO=(Re+d)= 6356.8+378521=384877.8 km

For speed of observer at Houston, we begin with speed of a point on a circle, calculate it as following:

V=2πR/time or duration

From Moonrise till Moonset, the observer spends 822 minutes for a half of a circle of radius Rp:

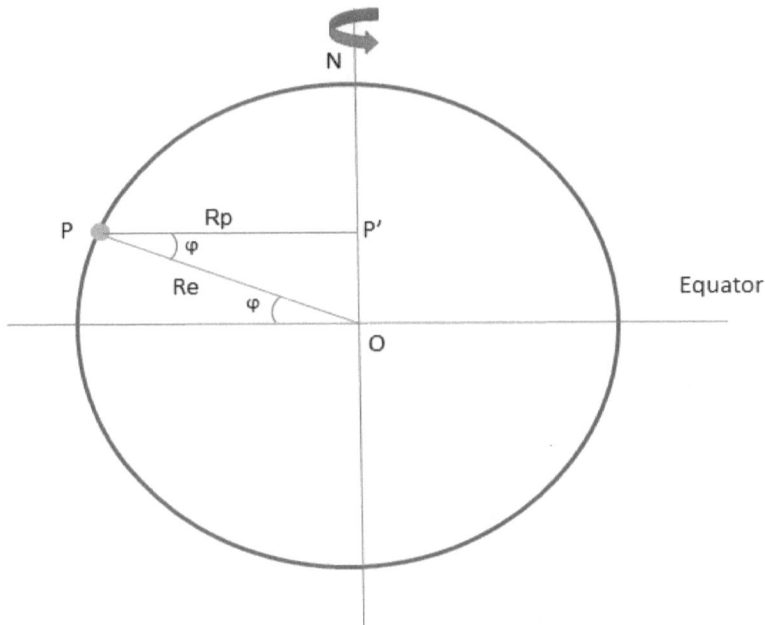

Figure 6/V-Latitude and speed of P while Earth rotating.

$Rp=Re*Cos(\phi)$, $\phi = 30°$, So we have
$Rp=6356.8*0.866=5504.9888$ km, Speed of P:

$Vp=\pi*5504.9888/822=$

$=21.03945$ km/minute $=350.65759 \approx 350.658$ m/s *(At ϕ =30° the Earth is rotating with 400.34 m/s, but in function with Moon's rise/set, it is 350.65759 m/s, b/c Moon's time indicates 822 minutes (14.20 hrs) from Moonrise to Moonset)*

For relative speed of Moon around the Earth, we apply the law of direct proportion as following:

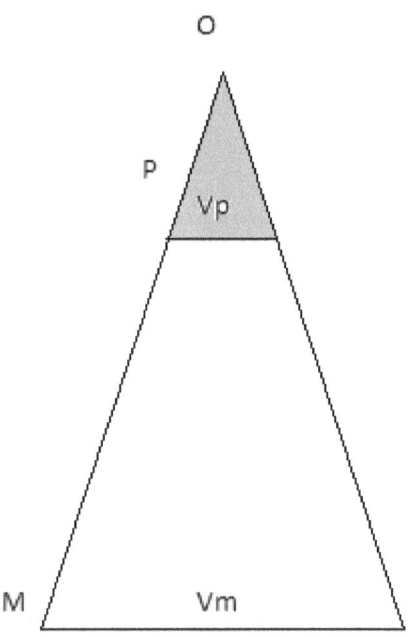

Figure 7/V-Speed triangle-direct proportional

OP/OM=Vp/Vm,

So Vm=Vp*OM/OP=350.658*384877.8/6356.8=

Vm=21230.8834 m/s

We have final calculation:

$|B(tc)|$=10^{-7}*Qm*21230.8834*378521^{-2}*10^{-6}*Cos^3(07.95^0)=1.43948*10^{-20}*Qm

3-Step 3: Better accuracy.

-Recall Biot-Savart expression for a single charge. We add some more lines on the basic figure 12b/v to facilitate the solution.

$$\vec{B} = \frac{\mu}{4\pi} * \frac{q * \vec{v} * \hat{r}}{r^2}$$

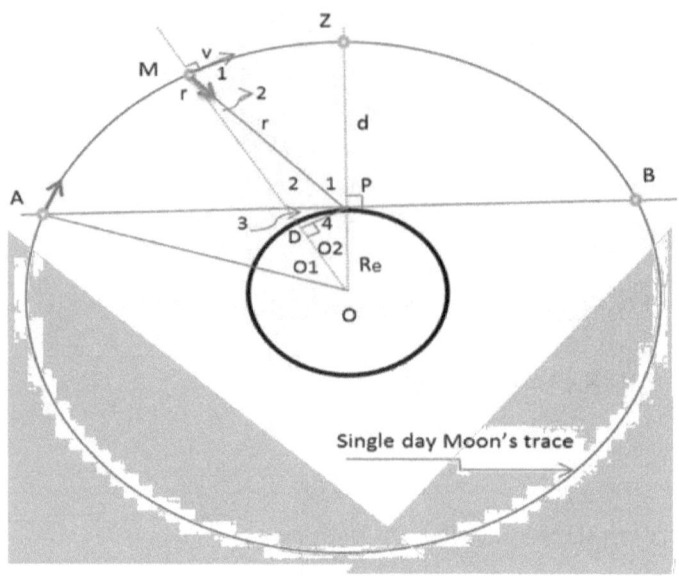

Figure 8/V-Moon's trace around Earth

- <u>In the above figure</u>: The relative Moon's trace (not the Moons' orbit) in 12 hrs is demonstrated (from Moonrise till Moon set) in 2D coordinates, the larger half of the circle (from A to B through under part) is not viewed. Within this duration, the Earth completes almost a half of cycle around its own axis. We can imagine that a certain observer standing somewhere near either North or South Pole, with a restricted view she/he can observes the Moon rising like depicted in the figure (while the Moon after Moonset is not seen, and the Earth makes no move).

The problem becomes pure geometric on 2D coordinates; and the figure requires some additional assumptions and quantities:

+ Moon is at a general position (M) on sky, its distance to P is "r" which varies against time and reaches the given value of d at meridian passing;

+ Moon is rising at A and setting at B;

+ Moon is expecting to pass meridian at Z over P; where $\widehat{APM}=\widehat{APZ}=90^0$ while Moon's altitude reaches its max of that day 75.7^0.

+ Vector $\vec{v}$ is perpendicular to the radius that drawn through Moon (OM);

+ PD is perpendicular to OM and parallel to $\vec{v}$;

+ The angles are named after each point and numbered: M1, M2 and O1,O2; or P1,P2,P3,P4;

+ Parameters: $d=378521*10^3 m$ (minimum distance from the given position P to Moon's centre).

+The Moon's angle (P2 or angle APM) is taken by a professional equipment (Sextant) (at the given time); (*the "altitude" of Moon or any object on sky is understood as the angle between the line from observer to the object and horizontal plane at observer. Therefore, a large gap between the two aforementioned concepts (altitude and P2) is found, and no chance for a mixing up or confusion*). Nonetheless, we can't deny that the Moon's altitude depends on P2 and reaches max almost at the middle of Moon's trace between A and B, therefore we may keep considering altitude in relation with Moon's angle P2. Furthermore, by the laws of 3D trigonometric, we can calculate P2 from altitude or vice versa, but no discussion for that issue is encouraged in this book.

The following is a figure that depicts approximately the altitude of Moon and the Moon's angle P2:

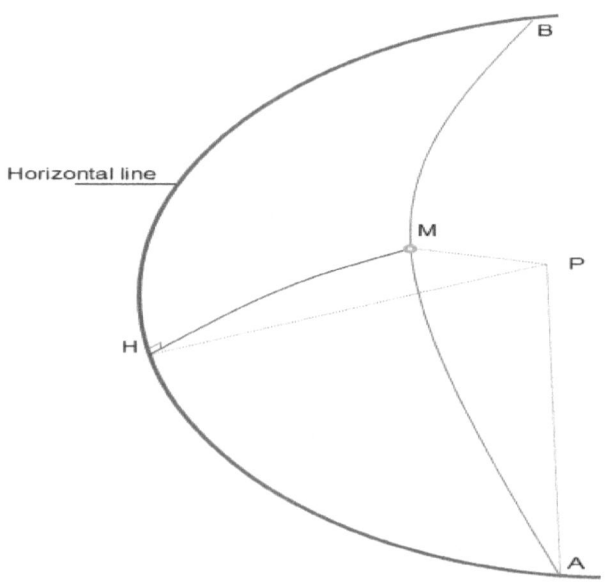

Figure 9/V-Moon's altitude

(At Moon's meridian passing, the angle APM is obviously 90^0 or $\frac{1}{2}\pi$ while Moon's altitude keeps varying (angle MPH is not always to be 90^0 even if we measure it from Earth equator)).

The following is the picture of equipment for measuring angle between the two objects or the altitude of Moon/star, the Sextant:

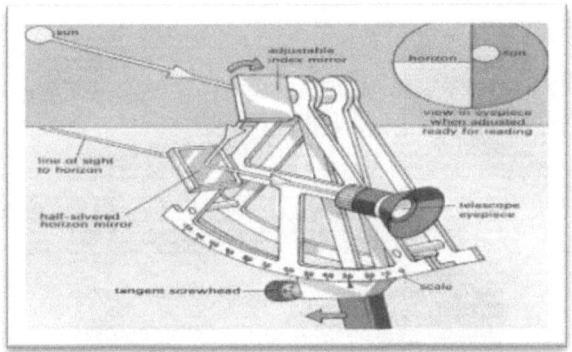

Picture 10/V-Professional Sextants

(The problem can't be solved without a professional Sextant, astronomy almanac (or Moon's time table), high-accuracy clock and calculator. This book is not for the discussion about the ways in using sextant; but we should bear in mind that for taking the angle from rising point to the Moon, we must set the sextant heading to the Moonrise's azimuth (A) then tilt & turn to (M)).

Again, we refer to the basic expression of Biot-Savart:

$$\vec{B} = \frac{\mu}{4\pi} * \frac{q * \vec{v} * \hat{r}}{r^2}$$

With Biot-Savart expression, this problem is the normal geometric one in astrophysics, we can demonstrate every important parameter on 2D coordinates. The sensitive issue in this expression is the product of vector $\vec{v}$ and unit vector $\hat{r}$

($\vec{v} * \hat{r}$). The product is a vector that directs out from paper toward readers, and its non-directional value is:

$$v*\text{Sin}(M1)$$

Where the M1 is the smaller angle among other larger ones between $\vec{v}$ and $\hat{r}$ at The non-directional value of B is:

$$B(tc) = \frac{\mu * Qm}{4\pi} * v * r^{-2} * \text{Sin}(M1).$$

Picture 11/V-The image of Astronomical Almanac.

(The above is image of an astronomy almanac by which we can find out some important data we need.

Entries are position (ϕ,λ) and date & time, the outlet gives parameters of the object we want, such as rise/set, meridian passing (with time, distance, illumination.

Although the information from Almanac is rather reliable but the real sextant reading is always encouraged, for reference and an alternative).

b/1- To seek for (r):

+ Consider the triangle MDP (Vector $\vec{v}$ and PD are extended for a better view with M'P'//MP):

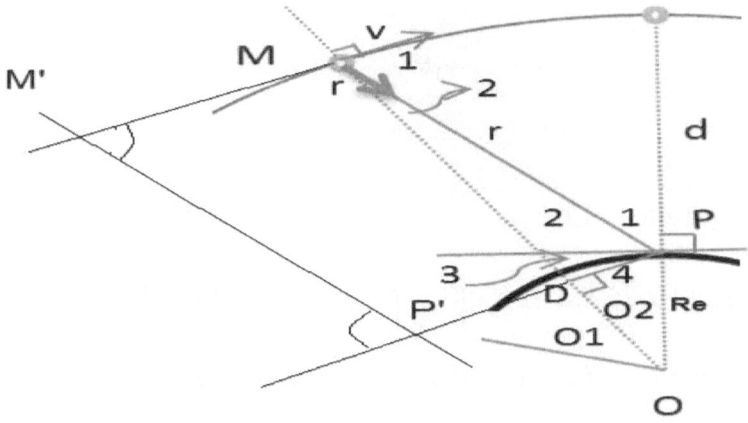

Figure 12/V-Triangle problems

The distance from M to P is "r" that varies against the angle at P of M and A

Note that the angle $\widehat{MPA}$ reaches π/2 when Moon is on its meridian, no matter how large the Moon's altitude $\widehat{MPH}$ is.

In triangle MDP, the MP (or "r") is hypotenuse and therefore:

$$MD = r*Sin(P2+P3)$$

The angle P2 is Moon's angle to rising point A, can be taken by sextant at 19.00 hrs. local time (but its value is given): $81.91°$ And P3 = O2, so we have:

$$r = MD/Sin(P2+P3) = MD/Sin(81.91°+07.95°) = MD/0.99999;$$

(Note: (Angle $\hat{O}2 = \frac{31*180}{702} = 07.95° = 07°55''$ (31 minutes before meridian passing and 702 minutes is duration from Moonrise to Moonset on that day);

$$Cos(\hat{O}2) = Cos(07.95°) = 0.990389$$

+ In the triangle MPO:

In the triangle ODP:

$OD = Re*Cos(\hat{O}2) = 6356.8*0.990389 = 6295.704795$ km

And $Sin(\hat{M}1) = Sin(81.91°+07.95°) = 0.99999$

$MD = MO-OD$ or $MD = (Re+d)-Re*Cos(\hat{O}2) =$
$6356.8+378521-6356.8*0.990389 =$

$= 378582.0952048$ Km

Thus, $r = MD/Sin(M1) = 378582.0952048/0.99999 = 378585.88106$ km=

$r = 378585.88106*10^3$m.

b/2-Angle M1:

M1 = P2+P3 (Vector v is parallel with PD, the rule for alternate angles is applied)

P2 is Sextant reading (or astronomy almanac), and the

P3=O2. Thus:

$\widehat{M1}$=P2+O2, with given P2=81.91⁰ and P3=07.95⁰

$\widehat{M1}$=81.91⁰+07.95⁰=<u>89.86⁰</u> = <u>89⁰51"</u>

Final calculation:

-Replace the followings in the expression and calculate value of $B(tc)$:

Radius r=378585.881*10³m

Speed v=21230.8834 m/s

Angle M1=<u>89⁰51"</u>, Sin(M1)=0.99997≈1.

$$B(tc) = \frac{\mu * Qm}{4\pi} * v * r^{-2} * Sin(M1)$$

As assumption of this problem, q=Qm Coulomb, every length is in meter (m), the unit of B is Tesla (T).

Permeability $\mu = \mu_0 = 4\pi * 10^{-7}$ T.m/A or (H/m), $\mu_0/4\pi = = 10^{-7}$

We have:

$B(tc)$ = 10⁻⁷*21230.8834*378585.881⁻²*10⁻⁶*Qm =

= 1.481287*10⁻²⁰*Qm Tesla

(The previous approximate calculation gives result:
*$B(tc)$=1.439480*10⁻²⁰*Qm)*

Because every of (r,v,M1) is depending on observer's

position, therefore the above result cannot be used for everywhere. Nonetheless it is to confirm that the Moon exerts its static electric force on the Earth and especially induces a magnetic field on it, the value can be calculated.

For instant, a reading on a Tesla meter is indicating that the Moon's contribution at Houston (South U.S.A) reaches its max $+5*10^{-7}$ T *(normally the meter reading about 2-3 hrs before meridian passing can indicate the average magnetic level)*; we can find Moon's total charge as following:

$B(tc) = 1.481287*10^{-20}*Qm = 5*10^{-7}$ Tesla

$Qm = B(tc) / 1.481287*10^{-20} = 5*10^{-7} / 1.481287*10^{-20} =$

$= 3.375443*10^{13}$ Coulombs.

(Problem: (the following problem is designated for 60 N to make sure that Moon signal is detected without being totally barred by plasma sphere): $3.375443*10^{13}$ Coulombs is a value of Moon's charge that we detected when semi-moon (illumination 50%) is passing meridian of a position "P" (60N, 0.0). If in 36 hrs later, when Moon's illumination of 40% we detect Moon's charge to be "0" at a certain position P' (60N, 180.0). Calculate the Moon's total charge components Qa & Qb?

This theoretic problem is definitely inspired by many science students but not included in this book. After this problem, we find out more about the roles of 2 hemispheres of Moon on E.M).

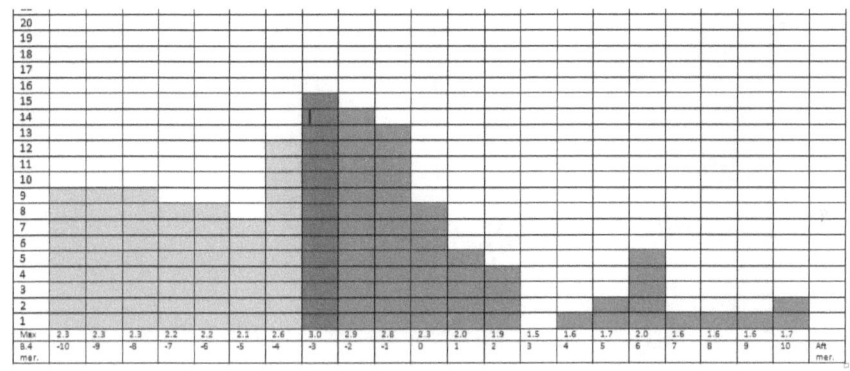

Figure 13/V-Earth's magnetic field amplitude (in mG)

The above is a real graph of magnetic amplitude observed by our correspondent at latitude 70 N on Atlantic, about new moon. The records are bold for 10 minutes before and after the Moon's meridian passing.

4-Problem3-Moon total charge ($B(tc)$) affects at Earth core:

A far-reaching debate about the magnetism at the Earth's core has been warm enough and cool. As matter of fact, in the temperature of thousands of C degrees, no real material is magnetized. But the magnetic field is not such heat limited, the magnetic field can be everywhere; the permeability is a coefficient that can indicate a limit but no link between temperature and permeability is discussed in this book.

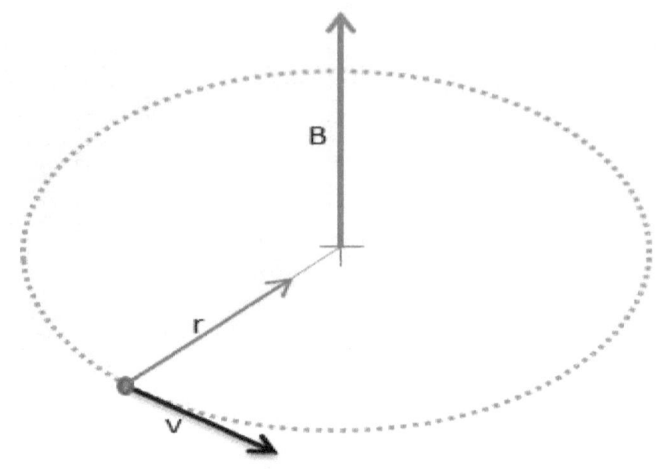

Figure 14/V-Earth's magnetism with Moon beside (repeated 8/v)

Moon is making a loop by which the Earth is inside, by both Moon's relative daily trace as well as the 24*29.5-hr day of Moon.

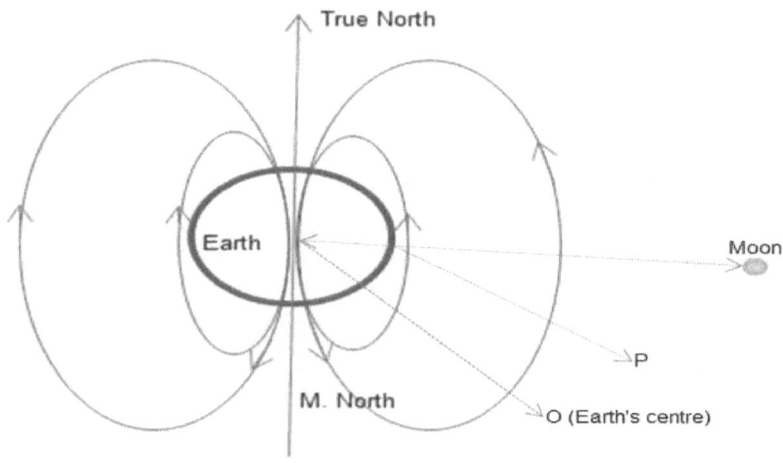

Figure 15/V-Field line of E.M is a loop

The dilemma we tackle with is that the Moon induces magnetism of the same direction to everywhere inside its loop (in figure 15/IV above), while the field lines of Earth magnetism are loops, each makes two directions; one is opposite to the other (the illustration is in the above figure).

For the Earth core, we should consider the average permeability of air (from Moon to Earth surface) and land material from Earth surface to Earth core. The result is grand total of a Δb(core) and the other is integral on hemi sphere of globe on Moon shined side; each of those two is opposite to the other.

This book does not offer such discussion, but mind that no field line is found in the Earth core if a Faraday's cage is covering the Earth.

4/1-<u>Two opposite points</u>:

We are working with the Biot-Savart expression 9.1.20 by which the direction of the magnetism induced by a moving charge is product of speed vector "v" and unit vector of radius "r" of the circle where the charge is orbiting:

$$\vec{v} * \hat{r}$$

Therefore, when we consider the influence of a charge "C" to a magnet, we encounter an argument as following:

The influence from moving charge is applying simultaneously on the whole of field line which makes a close-circle (figure 16/IV). From that figure, "C" influence is added up at P but contrary or diminishing to the magnet field at "P' ".

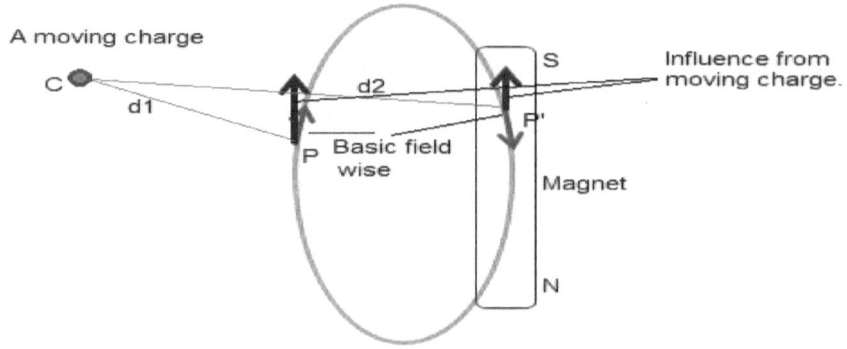

Figure 16/V- Basic and influence opposite.

On term of magnitude, the same expression also suggests the inversely square law by which the magnitude of influence is inversely proportional to distance from charge to the object (P and P'). So the influence at P is obviously much more than that at P'.

With the above mentioned inverse square law; as long as d1 is found shorter than d2, the influence at P is more than that at P', the real quantity is total of those two.

SUMMARY WORDS:

- <u>Huge electrified body under solar wind</u>: Although our calculation is almost done with assumption (or supposed figures), we got to recognize that the Moon is naked and independent, so it is a huge electrified body under solar wind while the Earth is inside the Van Allen shield and atmosphere; expecting to be less influenced by the Sun.

Under solar wind, the Moon's electric charge is varying and its change affects the Earth's magnetism, Biot-Savart expression 9-1-19 and 9-1-20 are applied. This is the matter of Moon's charge.

- <u>Moon facilitates detection to Sun's radiation</u>: we can detect the Moon's total charge, as indicated in some problem. After 2 consecutive semi-moons (2-week time), if we find out that the gap of Moon's charges between 2 times non-considerable; we may believe that the Sun's radiation doesn't change alarmingly, no matter how the Sun's flare could be capably seen from Earth. Otherwise we are alarmed with the change in the Sun's radiation. This is the obvious benefit that brought to us by the Moons' problem.

The mater of external influence is still going on, an argument that thought as sorted out totally is not really done completely; the axial magnetic vector is affected by the charge move, and this needs to be sorted out. How and how much the influence is? We discuss about it in the next.

CHAPTER IV: TWIN CONES AT POLES

I-Earth's Poles:
The Earth is moving around the Sun on the Ecliptic cum rotating with 24-hr cycle; its rotation is rather stable on ecliptic and defines 2 geographic poles. The straight line connecting the two poles is named as Earth's own rotation axis.

This chapter is a discussion about the influence of externality to the areas adjacent to the poles and especially the magnetic fields in there. And more than that, it is introducing a new concept about the complicated magnetic field at pole-the "Polar Fields" as I name them so, which are created by the charge(s) on rotating move around the rotation axis.

Meantime, note that we restrict the scope of our discussion on the external areas around Earth rotation axis and none for the internal areas.

Ambiguously, we should note that the Earth is polarized with North and South magnetic poles which are found as not coinciding on Earth's geographic poles, besides they are likely North-side-South.

II-Twin Cones at Poles:

One of previous problem is repeated here:

If total charge of the Van Allen belts and plasma sphere is equivalent to a single band of $+10^6$ Coulombs around Earth equator at a distance 1500 km from Earth surface, (we name the single band "the Band"), and evenly distributed on it; if the Earth makes no move on its orbit throughout 24 hrs.

<u>Question</u>: "If each angle at Z is $\pi/3$ (Figure 1/v). How is the magnetic field that induced by the Band to the point Pn or Ps on rotation axis of the Earth?"

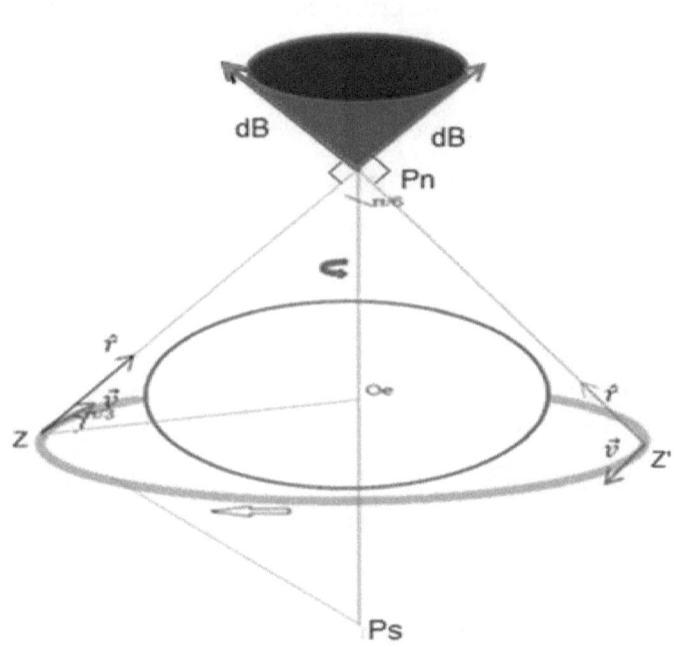

Figure 1/VI- Field lines at Pn on rotation axis

This problem is solved in the manual book where Biot-Savart theory is applied for several moving charges; and its simulation is as following:

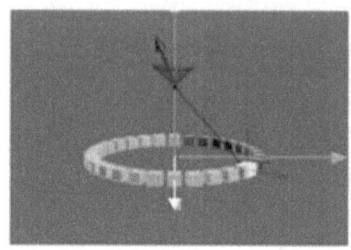

A ShockWave simulation of the use of the principle of superposition to find the magnetic field due to 30 moving charges moving in a circle at an observation point on the axis of the circle.

Figure 2/VI- The 30 charges moving on a circle (C.R University lab.)

Thus, we need just solve our problem in format of the above example for the single band; then every similar problem for charged point on the Earth can be solved by the similar way.

What normally requires a particular attention is the wise of dB or product of 2 vectors $\vec{v} * \hat{r}$. At the poles the field lines make cones, one under the other adjacent to Earth's rotation axis; the sexuality of the moving charge decides the sexuality of the cone it weaves which might either head up or down.

<u>Our problem in brief</u>:

- The Band is around the Earth at 1500 km altitude above

Earth's equator and charged with $+1*10^6$ Coulombs.

- The two points on Earth's rotation axis, one is at the North (Pn) and another is at South (Ps) and they are both at the same altitude above Earth surface.

- Everywhere on the Band is charged evenly as assumption, and so the figure is simplified as following:

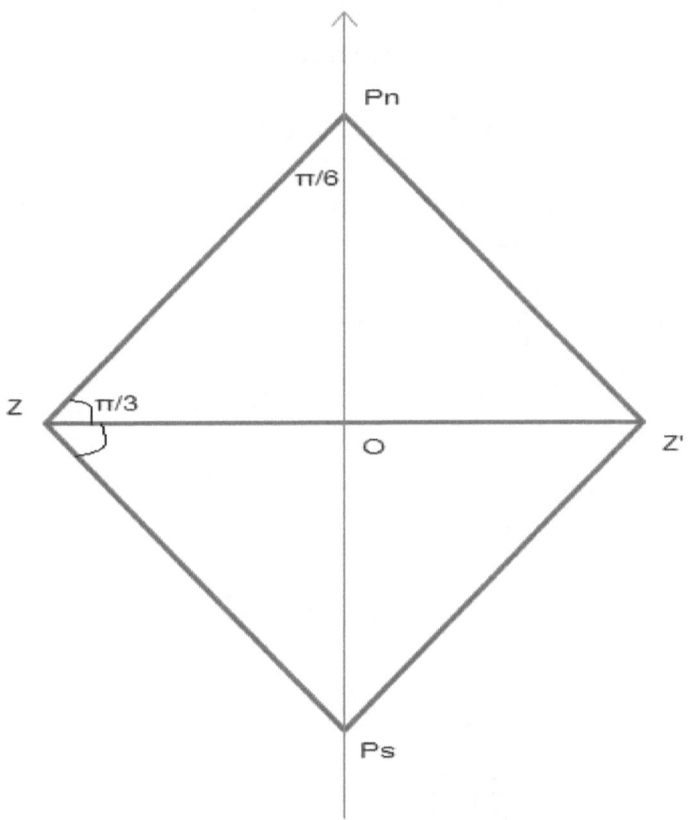

Figure 3/VI- Influence of the Band at Pn and Ps

Solution:

OZ = OZ' is distance from the single band to Earth's centre and is:

OZ = 6356.8 + 1500 = 7856.8 km = 7856.8×10^3 m

In general figure, ZPn is hypotenuse of the right triangle ZOPn:

ZPn = OZ/cos(PnZO) and with PnZO = $\pi/3$, we have:

ZPn = OZ/cos($\pi/3$) = $7856.8 \times 10^3 \times 2$ = $\underline{15713.6 \times 10^3}$ m

- This "ZPn" is the figure to be replaced to "r" in Biot-Savart expression 9.1.20

We call back the parameters in the last problems:

$$|\vec{v} * \hat{r}| = v*\sin(90°) = v$$

For q*v in Biot-Savart expression 9.1.20, we treat it as a module, in general:

$v*q(i) = (2\pi*R/24*60*60)*(q(tc)/(2\pi*R))$ = $q(tc)/(24*60*60)$

And the result for each dB is:

dB = 10^{-7} * $q(tc)/((24*60*60)*(15713.6 \times 10^3)^2)$

With total charge q(tc) = 1×10^6 Coulombs, we have dB:

dB = $1/((240*60*60)*(15713.6 \times 10^3)^2)$ =

= $1/(864 \times 10^3 \times 10^6 \times 15713.6^2)$

And so, the non-directional value of dB at Pn or Ps with $+10^6$ Coulombs band is calculated by (9.1.20) is:

$dB = 4.68743*10^{-21}$ Tesla

The real band charge is varying at all time and not as tiny as 10^6 Coulombs as assumed for this problem, hope that NASA can figure out the Band's properties one day.

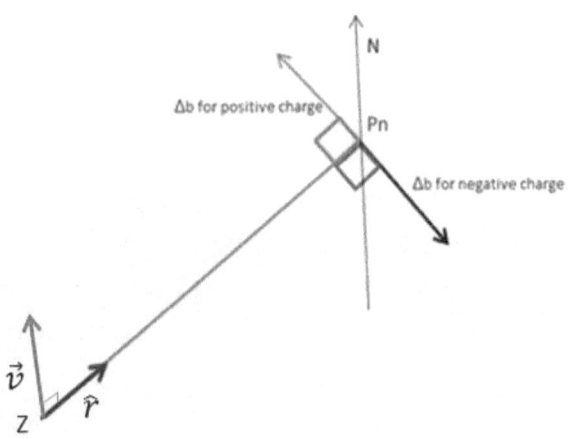

Figure 4/VI-The twin-cones establishment

Conclusion:

-By vector rule, every Δb is perpendicular to either ZPn or ZPs and the vector v ($\vec{v}$) at respective Z(i).

-By Biot-Savart, every charge placed adjacent to the pole can induce magnetism to it and rotation axis. Each charge can weave a cone within 24 hrs.

III-The Polar Fields: We used to consider and imagine the E.M field lines like the following images:

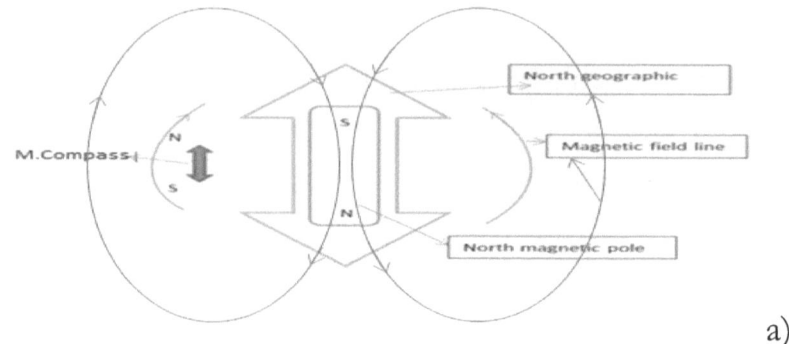

a)

For "0" deviation or:

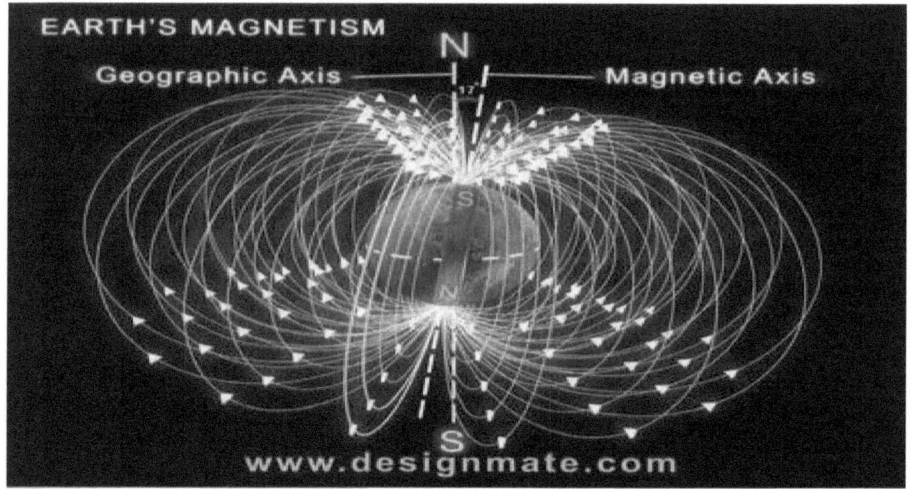

b)

Figure 5(a+b)/VI-Normal E.M field lines in magnetosphere

The above diagrams are always right with a magnetosphere but not a real complete image of E.M. And we never thought of the idea of twin cones until this discussion, the twin-cones are created by charges during Earth's rotation and surrounding its rotation axis (Geographic Axis).

1-Cones on the axis:

The Band (supposed as $+10^6$ Coulombs) is relatively moving around the Earth at all time, therefore the field lines it induced are weaving cones (Figure 1/V). The (-) Band (if any) does induce the field lines of cones that opposite to ones of the (+) Band.

-The important note for a cone (can be produced in lab) is: The further from pole, the more obtuse the cone is.

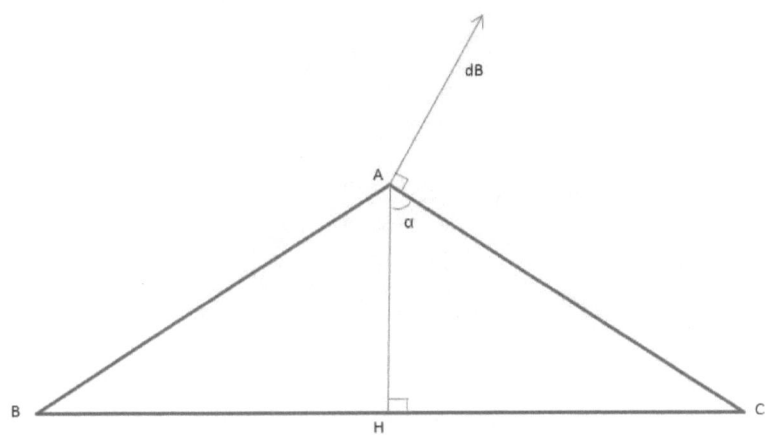

Figure 6/VI-Direction of dB

This argument is explained by the value of Tan(HAC) or Tan(α)=HC/HA; the value of this function becomes "nil" when HA is too much more then HC, the angle "α" then becomes so tiny that the vector dB might be viewed as parallel to HC.

- The important note for this twin is: the field line variety.

+ The field line induced by a negative charge is opposite to one of positive charge (Δa vs Δb in Figure 4/V).

+ It is not continuous as one of a magnet, but both do either start or stop at rotation axis. The following figure (Figure 9/V) describes the twin head-to-head opposite cones:

Figure 7/VI-Field lines making twin cones (not drawn in scale)

+ Meantime one looks aligned to the other but actually although both two cross the same point on rotation axis and one is aligned to the other; each acts on a direction that opposite to the other.

+ Especially each acts on a different time or no simultaneity is found. That's why each one never expects to eliminate the other in a twin-cone.

Every two opposite charges on the same latitude make a twin-cone on Earth rotation axis above pole as demonstrated in the Figure 8/V.

Many people come with a question that whether we can count or guess how many cones are there? Definitely we fail to give out a number but affirm that "lots and uncountable".

The following figure is presenting 2 twin-cone pairs at 2 different points on Earth's rotation axis; each of the cone pairs is induced by a pair of opposite charges fixed at the same point (P) on Earth. The point (P') is another position of (P) at another time of a 24-hr cycle.

The field vector is to be perpendicular to PZ. Therefore it should be said again as: the further from pole, the more obtuse twin-cone is; and that the closer to the pole, the sharper the cone is.

So the twin becomes one plate of 2 layers incorporated where each vector looks as eliminating the other when they are far from pole and become one likely. Nonetheless, the author of this book does not confirm the mutual elimination.

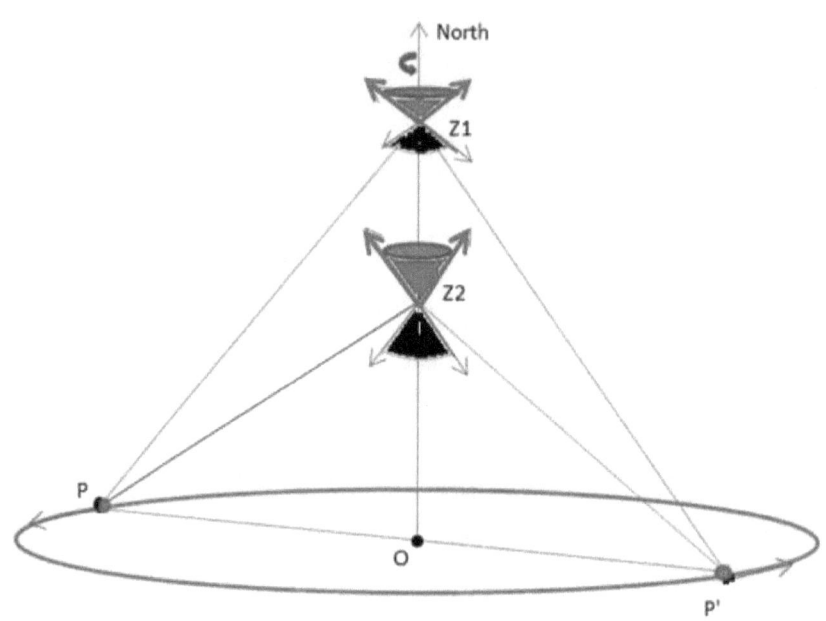

Figure 8/VI-Two twins on axis

2-Cones off the axis:
As soon as getting off the axis, the distances between the observer and the charged band are to change or KP-KP'≠ 0. We get to discuss about the change in accordance with the position of observer off the axis.

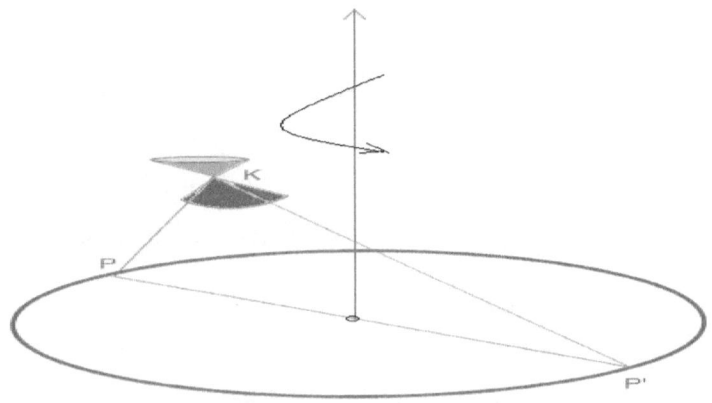

Figure 9/VI-Field lines at K off Axis

For a certain "K" off the axis, the distance from K to a nearest charged spot (P) on a band is KP, while the farthest spot is P' and the distance is KP'.

As KP'>KP, by Biot-Savart expression 9.1.20, the field line twin-cone looks like illustration above in Figure 8/V.

In short, there are lots of cones and twin-cones woven by field lines found above poles and around a charged band. The further from Earth rotation axis, the more deformed the cone/twin-cone is. Every charge on Earth can draw a charged band within 24 hrs, and make a cone in pole's space. This concept draws an image that totally different to one of a magnet field lines.

There is more than 1 opinion, those assume that we can sum up or totalize the vectors of each cone to produce one total vector. This can be done in a lab but rarely found in reality because we almost rarely find one vector member being created at the same time with another vector member of the same cone.

On the other hand, every real charge on Earth can be on-and-off in a short life; the charges of thunder cloud and its counter patch on land are brilliant examples. Thence the polar field can be affected accordingly and in such an unstable condition.

Even the so-called cones that induced by the Band, everyone almost can't be completed as a real cone before the charge disappears or moves away from its nominated orbit. If consider the life of each thunder cloud as 4 hrs, a "fan" of 1/6 cone is established at a certain altitude on Earth Axis:

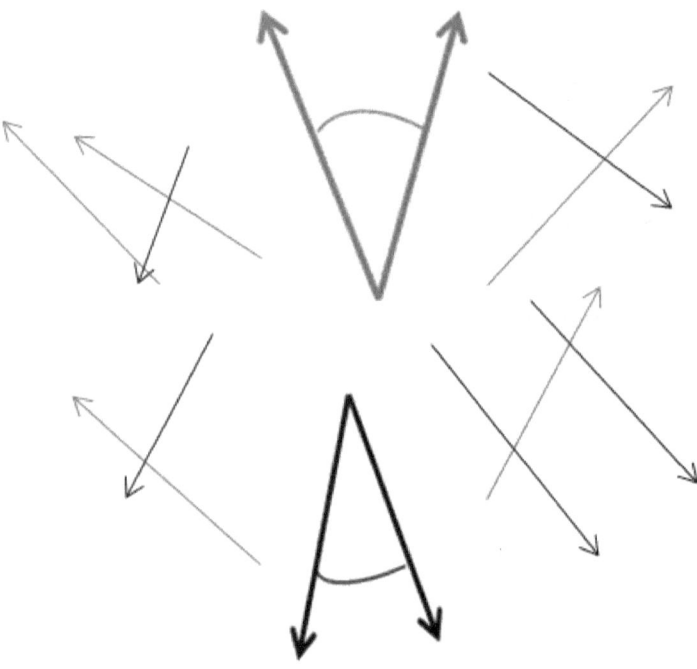

Figure-10/VI- The field vectors at Polar Field

Some vector is even so thin as an arrow. All of them make the magnetic field at the Earth's geographic poles, a huge pile of vectors of all kinds and all types, adjacent to magnetic pole; that is Polar Field. The Polar Field is established in electric field of charge(s), at the poles and mixed in the normal field of magnetosphere of every rotating object.

The above image partly illustrates the complicated properties of the Polar Field. As known to everyone, a vector quantity always consists: how large, what direction, what wise and where does it apply to?

The question "where does it apply?" is most difficult one with E.M because the Earth never stops.

(Problem: A couple of thunder cloud at 550 km altitude and its counter patch on land at Houston (30N, 95W), given charges: $+10^6$ and -10^6 Coulombs, given duration: 4 hrs. Assume that the air permeability $\mu=\mu_0$.

Question: How is magnetism induced by those 2 charges at North magnetic pole on 450 km height?-This problem is extra and purely theoretical, not solved in this book.)

III-Polar Fields at North and South:

1-Source for Polar Field at North Pole:

In addition to Van Allen belt and plasma sphere, the charges on air, on seas and land can contribute to magnetic field at Pole. The followings are 2 images of thunderstorm and lightning of Europe and U.S taken in July 2017 for further consideration.

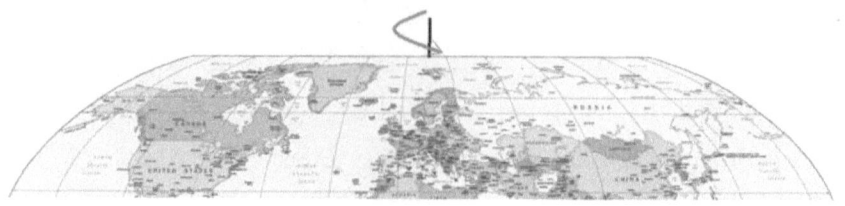

Figure 11/VI-The chunk of sea & land charges at North

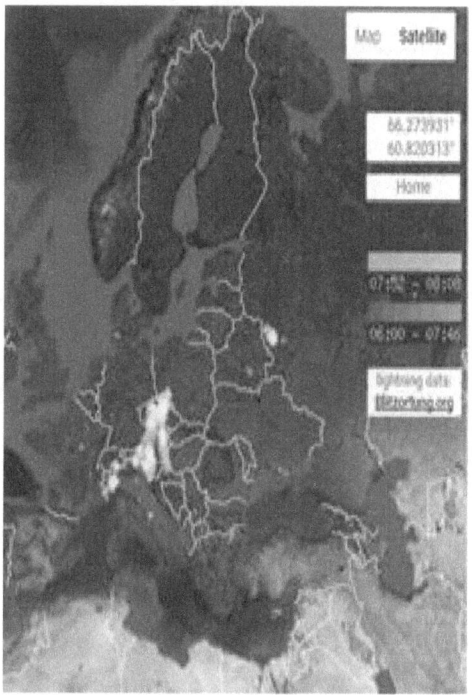

Figure 12/VI-Thunderstorm at Europe

We do not set and solve any problem for a certain charge on Earth because the result makes no difference to the above, but we do affirm that the mass of charges on Earth can draw lots of bands and each of them induces uncountable cones.

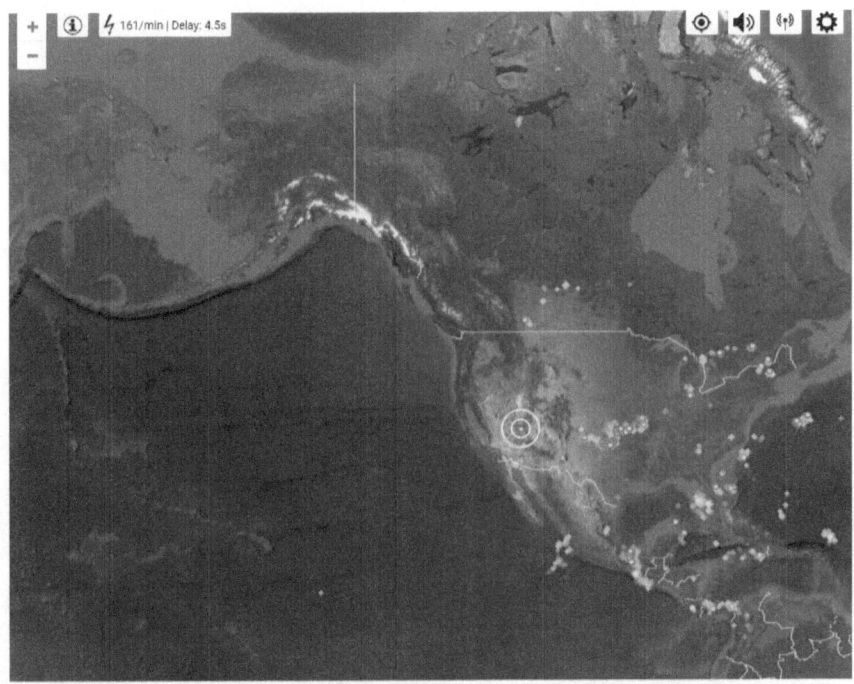

Figure 13/VI-Image of thunderstorm in North American Continent.

Right after establishment, each thunder cloud together with its counter patch on land definitely contribute to Polar Field until disappearance.

2-Polar Field at South Pole:
Like the North Pole, the South Pole is under Polar Field which is induced partly by the Band. But unlike North Pole, the South Pole is among a purer community of charges on land.

The following image is a chunk of the South Pole and somewhere adjacent to it.

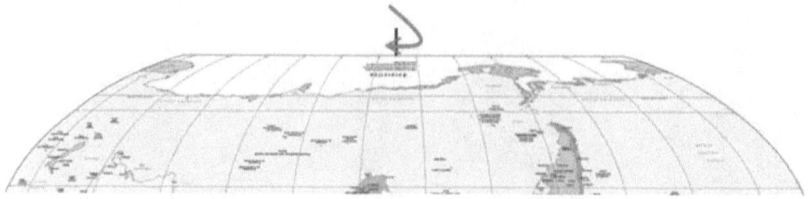

Figure 14/VI-A chunk of seas and land at South

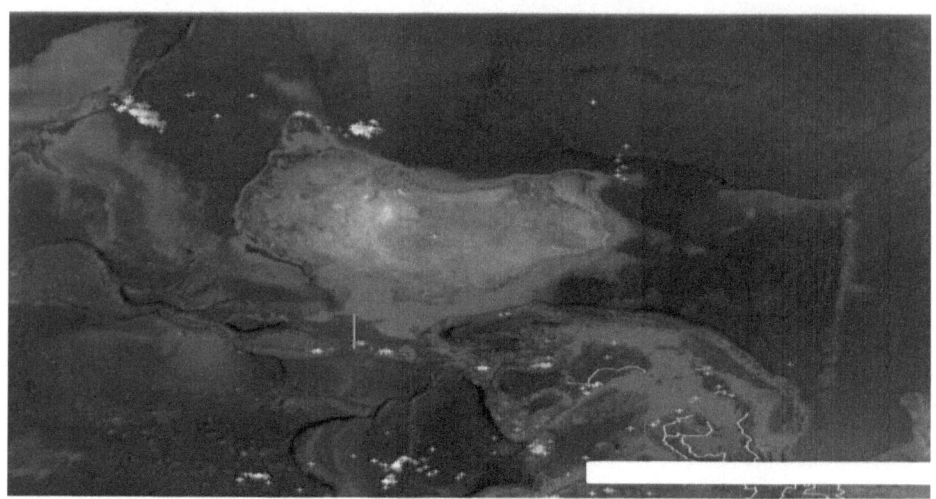

Figure 15/VI-Thunderstorm at South.

The Figure 15/V is exposing some thunderstorms around New Zealand, South Australia which definitely contribute to Polar Field at South. The charge mass in there looks much poorer than that of the North. The poorer condition, as such I believe, definitely should make purer Polar Field.

IV-Light intersection near poles and pole light: Aurora.

Figure 16/VI-Aurora seen from ISS

Nowadays, the Aurora is quite known to everyone. This is a phenomenon that happens only at poles.

As mentioned before that the scope of this discussion is only for outside or external to the Earth. We now should revise our scope that we do not go beyond magnetism; therefore Aurora is only an extra discussion that mostly quotes excerpt from other researchers.

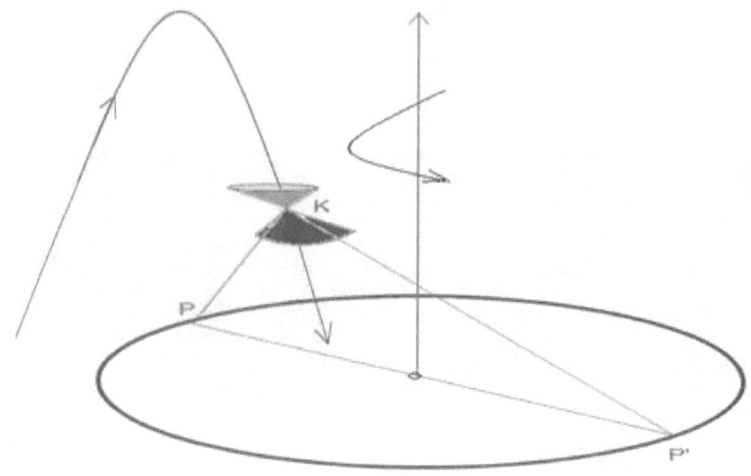

Figure 17/VI-E.M field line and cone

On the other hand, the above figure is illustrating the intersection between E.M field line and an off-axis cone adjacent to Earth magnetic pole. It is not hard to point out a total vector of a cone member vector and E.M vector, but it is quite very hard to say how the field affects the charged particles.

Notwithstanding, as long as the phenomenon happens in the Polar Field, we are titled to presume that the Polar Field contribute either partly or considerably to that happening.

In general about Aurora, we can't entertain ourselves

without reading the thesis or essay for Aurora title.

The following is some excerpt from wikipedia dictionary:

An **aurora**, sometimes referred to as a **polar lights** or **northern lights**, is a natural light display in the sky, predominantly seen in the high latitude (Arctic and Antarctic) regions.[a] Auroras are produced when the magnetosphere is sufficiently disturbed by the solar wind that the trajectories of charged particles in both solar wind and magnetospheric plasma, mainly in the form of electrons and protons, precipitate them into the upper atmosphere (thermosphere/exosphere), where their energy is lost. The resulting ionization and excitation of atmospheric constituents emits light of varying color and complexity. The form of the aurora, occurring within bands around both polar regions, is also dependent on the amount of acceleration imparted to the precipitating particles. Precipitating protons generally produce optical emissions as incident hydrogen atoms after gaining electrons from the atmosphere. Proton auroras are usually observed at lower latitudes.[2]

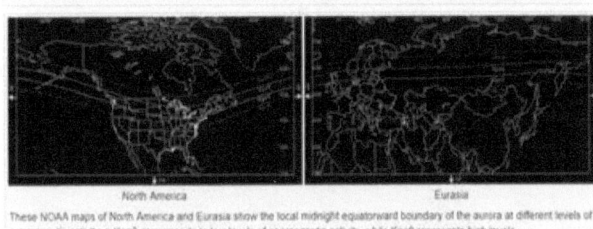

North America   Eurasia

These NOAA maps of North America and Eurasia show the local midnight equatorward boundary of the aurora at different levels of geomagnetic activity; a Kp=3 corresponds to low levels of geomagnetic activity, while Kp=9 represents high levels

Auroras are occasionally seen in latitudes below the auroral zone, when a geomagnetic storm temporarily enlarges the auroral oval. Large geomagnetic storms most common during the peak of the eleven-year sunspot cycle or during the three years after the peak.[10][11] An aurora may appear overhead as a "corona" of

...

"Auroras are occasionally seen in latitudes below the auroral zone, when a geomagnetic storm temporarily enlarges the auroral oval. Larger geomagnetic storms are most common during the peak of eleven year sun-spot cycle or during the three years after peak. An aurora may appear over head like a corona of rays, radiating from a distant and apparent central location, which results from perspective. An electron spiral (gyrates) about a field line at an angle that is determined by its velocity vectors, parallel and perpendicular, respectively to the local geomagnetic field vector B. This angle is known as "peak angle" of the particle. The distance, or radius, of the electron from the field line at any time is known as its Larmor radius. The pitch angle increases as the electron travels to a region of greater field

strength near the atmosphere. Thus it is possible for some particles to return, or mirror, if the angle becomes 90 degrees before entering the atmosphere to collide with the denser molecules in there. Other particles that do not mirror enter the atmosphere and contribute to the auroral display over a range of altitudes. Other types of auroras have been observed from space e.g. "poleward arc" stretching sunward across the polar cap, the related "theta aurora", "dayside arc" near noon. These are relatively infrequent and poorly understood. There are other interesting effects such as flickering aurora, "back aurora" and sub-visual red arcs. In addition to all these, a weak glow (often deep red) observed around the two polar cusps, the field lines separating the ones that close through the Earth from those that are swept into the tail and close remotely. "

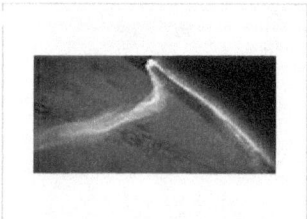 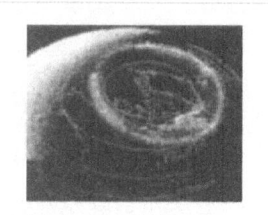

Aurora during a geomagnetic storm

Diffuse aurora observed by DE-1 satellite from high Earth orbit

Figure 18/VI-Aurora images

We can find many essays from some other authors, every of them almost explain how and why. However, each author can entertain us with a different emotion.

CONCLUSIONS

After all the problems in this book, I trust that we should re-consider our assumption about the Earth Magnetism in general.

My conclusion in brief:

1-External factors: We did assume that they all evaluate about 5%-6% of total E.M.

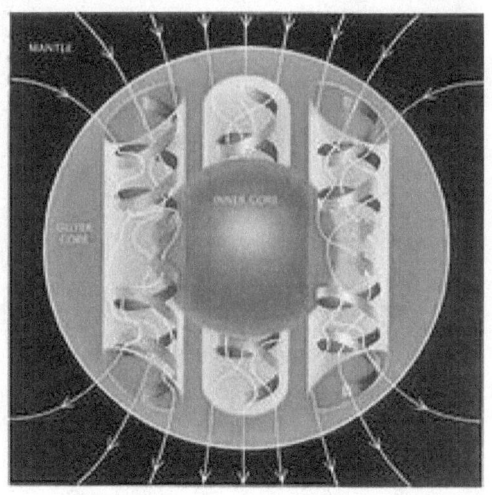

Figure 19/VI- Rotors in the Earth

But after solution to several problems, the external factors (including sea & land charges) who contribute to hourly variableness such as the sea-land charge variableness likely; are those who dominate the change of the Earth

Magnetism. For the daily/monthly variableness in Earth's Magnetism, it is rather depending on Moon's position. Furthermore, with Van Allen belt and the Moon as brokers, the Sun seems to influence a lot on Earth, the magnetism is considered specifically.

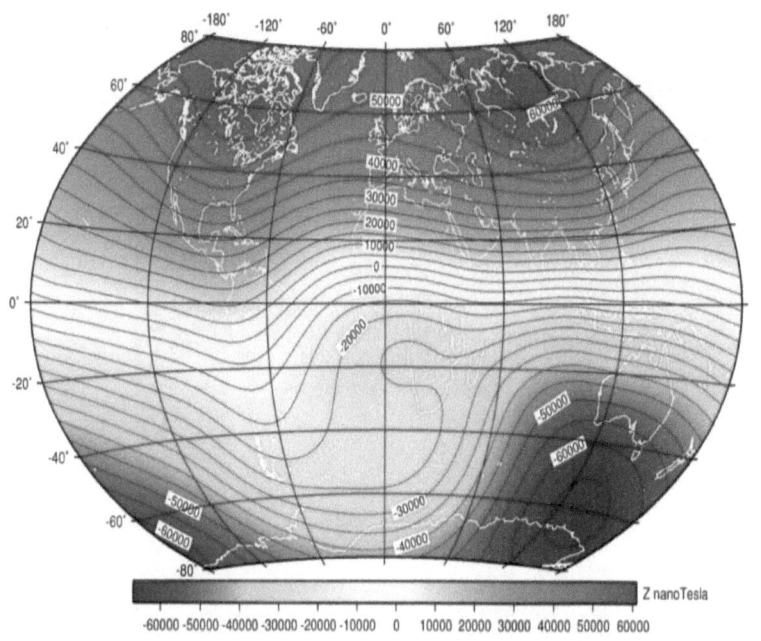

Figure 20/VI-Vertical intensity of Earth magnetism, recorded in 2015

The above image is the latest survey report from British

scientists which depicts the arrangement of Earth magnetism that recorded in 2015.

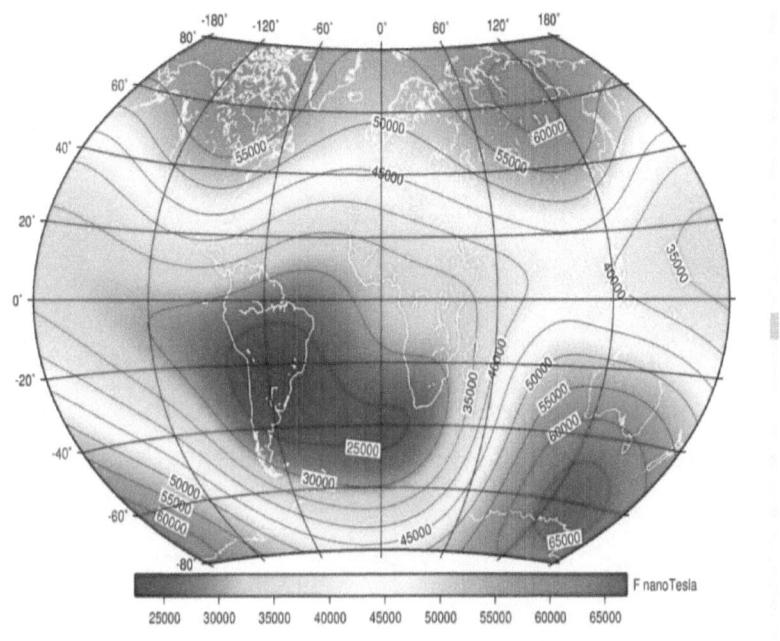

Figure 21/VI-Horizontal intensity of Earth magnetism-2015

2-Land charge and lightning: This is considered in chapter II of part I. After the problem about land's contribution to E.M, we find out that the positive charge on land diminishes E.M.

The positive charge on land analysed in that chapter is on area under thunderstorm. The law of Lenz is working right after freeing of millions of Coulombs (by the lightning which makes a huge jolt in E.M), that's why we don't just find "jolt" but also "echo of jolt" separately (a short retardation).

3-The sea surface charge: This is a part of chapter II of part I. The sea surface is assumed as negative, in fact there is less convection on there and certainly the sea surface is to cope with the change of land charge to maintain a neutral Earth, but the question is:

Late?

The retardation is suggested and mentioned right in the mechanism of coping by which the ion sheath is staying longer on sea surface. Note that the ion sheath is a vulnerable layer.

Limit?

Both ion layer thickness and life time are definitely limited; the convection at sea is not so tense as on land but it works, so the ion sheath is not built up to the sky. On the other hand, the lateness does not mean "forever". Therefore the sea-land balance is vulnerable and can be broken in favour of either sea or land occasionally.

Hourly variableness?

Due to the change of air-earth conduction current, the land charges and sea surface charge make E.M varying in accordance with the change of thunder storm area, which in turn depends on relative location of Sun to the Earth.

4-Polar Fields: This new concept is introduced in chapter IV/ part II. The Van Allen belt and plasma sphere, the charges on land as well as at seas near a pole can produce field vectors of all kind, all types. The pile of vectors is named: the Polar Field; which can be found at every pole of any rotating object with charges around.

5-Industrialization and E.M: As mentioned above, ion sheath is vulnerable and its life is quite restricted. Another question is whether the industrialization in recent centuries is the major reason for the decline of E.M amplitude? The ratio of positive ions to negative ions in each CC of air in city and industrial area seems to confirm what we assume in this conclusion.

6- Inertial surge: We assume that the sea is one that can stand up to the land on term of electric charge. The figure (6/II) is to point out that the thunderstorm area keeps changing at all the time or every minute. This is another note about E.M which likely is diminished by thunderstorm area, or we can say: the larger thunderstorm area, the weaker E.M is. In return, the E.M must surge right away after a lightning strike which may free millions of positive Coulombs from Earth surface.

7-Six main compositions of magnetism:

Although several compositions of Earth Magnetism still require further debate, about both quantity and quality; although each composition of E.M is varying at all the time; I do display desperately all of proposed compositions for a general view about E.M.

The following is figure of 6 compositions; individuals and then each is stacked over the other:

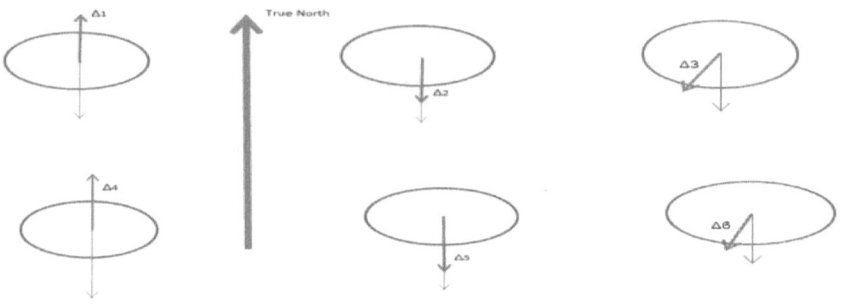

Figure 22/VI-The single compositions of E.M-the $\Delta(i)$

+ $\Delta 1$ in chapter II of part I, its value is obviously unknown and certainly not estimated desperately. Nonetheless, it is proved to be totally contrarian to current E.M.

+ $\Delta 2$ is presented in chapter II of part I, proved to be backing E.M.

+ $\Delta 3$ is in the chapter I of part II, which presents the influence of the Belts and the plasmasphere to the E.M. In fact there never been a fixed value or definite shape of the Belt found because the so-called value and shape are always varying in accordance with the solar activity as well as relative position of one to the others. Although the result requires more debate, I do desperately attribute it a wise or sexuality which is backing E.M.

But note that its direction is not coinciding with Earth's rotation axis which tilts approximately 23.5 on the Ecliptic.

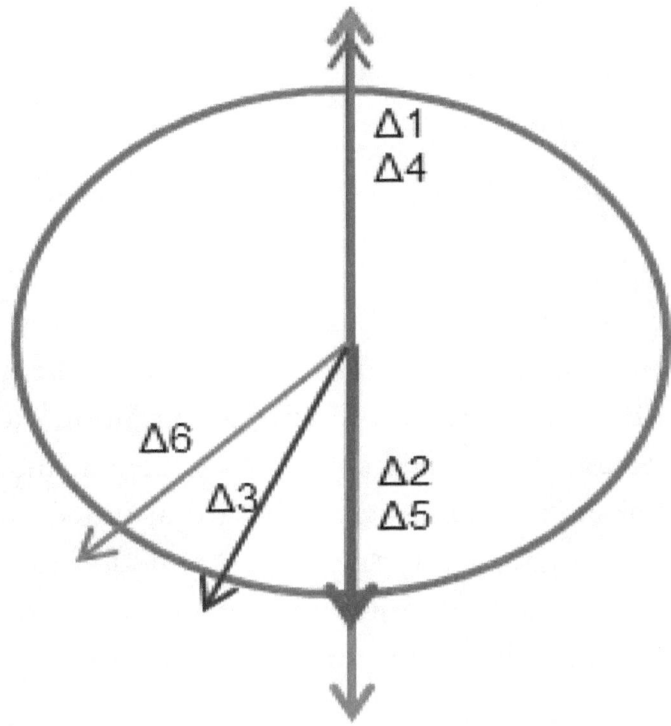

Figure 23/VI-Every Δ delta stacked on the others.

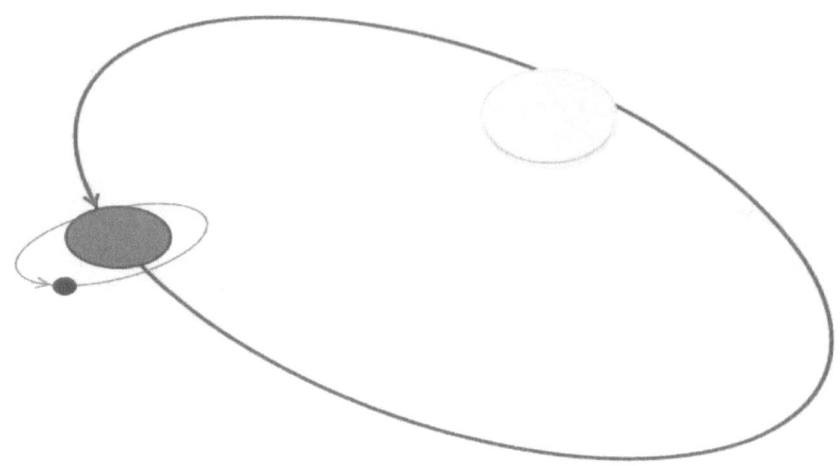

Figure 24/VI-Sun-Earth-Moon on sky

+ Δ4 and Δ5 -The Moon: Actually at any time there must be a certain value of total Lunar charge, and on top of Lunar total charge is a sine function of Moon's hemispheric charge. But there is no information about Moon's total charge found; we have no choice but to assume the Moon as neutral and suppose a calculation for its influence to E.M. Moon's day of approximate 27*24 hours is supposed as 2 halves, each is opposite to the other on term of electric charge on there, hence the influence to the Earth.

These two are calculated with assumption that the Moon is on no move, therefore each wise is to be on the Earth's rotation axis.

+ Δ6-The total charge of Moon is a huge number although it is varying at all the time. However, as long as the assumed figure is not verified, we just give desperately a number and try a mock calculation. The only confirmation is that its influence to E.M must be so powerful that together with the Belt's influence cum undefined rotor to make Earth's magnetic pole deviated away from geographic pole.

8-Local influence:

Although every mentioned factor contributes to the deviation of the magnetic poles, but we must be mistaken if ignore the local influence. Earth Magnetism has been confirmed as influenced by external bodies and varies at all the time or hourly; actually it also depends on characteristics of local rock and soil.

This argument of local dominance is recognized at once if we suppose that a certain rotor is laid on Earth's rotation axis, a strong magnet is laid in U.S and Canada to set a pole at North Canada so that the total magnetic vector is now 2000km from True North toward Canada, where we call "the magnetic pole". This book is not about the localism or the magnets of the Earth, but the author won't oppose if somebody wants to do so.

As matter of fact, if a magnetic compass in the field of a magnet, it definitely bears an error that we can't ascribe the E.M rotor.

The movement under Pole areas can be another issue that undisputed, the North and South magnetic pole can be affected accordingly, and so again mind that we can't ascribe every move to the E.M rotor.

9-E.M Flip: Nowadays, E.M flipping is a hot topic, someone even point out that the North magnetic pole is drifting from Canada toward Russia, so it is supposed to be flipping in thousand-year time.

Under influence of solar wind, the Moon keeps reverting within every fortnight (about 2*7 days) and the positive charges on land keep diminishing E.M; while the Van Allen belt and sea surface keep maintaining Earth Magnetism against the reverting. I speculate that if the sea loses charge until it is balanced then less to the land on term of magnetic influence, the E.M may be at the risk of flipping which depends almost on the change in cosmos and rotors inside the Earth.

10-A big sum: A question that we can't dodge an answer is: What are compositions of a magnetic vector at any locale on the Earth? The answer must be "7", whereby six compositions are mentioned above and another is $\Delta 7$ or $B(e)$, which is the composition that pre-set and available in the local soil or stone that normally amounts about 10% or less of total intensity. The exception is somewhere near power cables, industrial area or other locale sources; otherwise the composition is maintained by undefined rotors or sources inside the Earth or locale. All 7 vectors make one total vector on magnetic compass, this is applicable everywhere and makes X (Northern wise), Y(Eastern wise) and Z(vertical wise) compositions in global coordinates:

$$\vec{B} = \sum_{i=1}^{7} \Delta(i) = \sqrt{X^2 + Y^2 + Z^2}$$

As soon as we find out the function for B, we can apply the **Magnetic Induction Equation (MIE)**:

$$\frac{\partial \mathbf{B}}{\partial t} = \eta \nabla^2 \mathbf{B} + \nabla \times (\mathbf{u} \times \mathbf{B})$$

or

$$\frac{\partial \mathbf{B}}{\partial t} = \eta * \nabla^2 * \mathbf{B} + \nabla * (u * B)$$

Nonetheless, perhaps the function for B will never be established (the influence from cosmos is quite non-anticipated then the Laplace operator can't be "nil").

The unanswered question can be at everywhere, especially at Δ7 which is depending not just on intensity of undefined rotor but also on local magnetic characteristics.

Thus, relatively when we consider one from seven, that one is the odd composition to the rest of Earth Magnetism, and the rest as mentioned is to make Earth Magnetism sphere. The odd composition can only be determined if the rest of Earth Magnetism has been confirmed in advance.

11-Unsolved problems:

Earth Magnetism is a giant problem of Earth Science, my

ambition is to contribute a bit to this subject and trust that I do enough on my own as an individual. The far way to destination should be left for many Earth lovers to join.

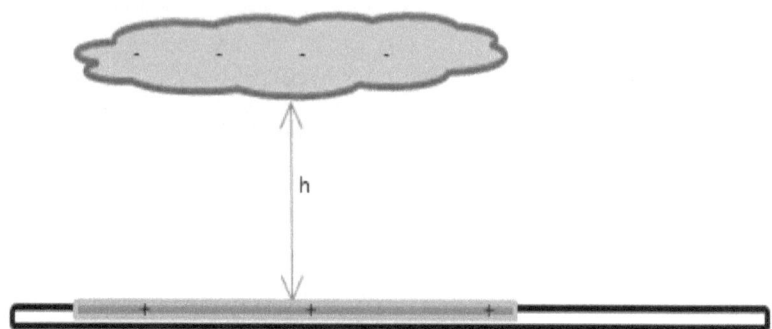

Figure 25/VI-Cloud patch above charged area

- The casual influence from every charged cloud patch is the first I leave back. The charged cloud (including those in thunder storm) is obviously to influence on E.M; nonetheless neither cloud data nor the argument is not always available for "set & solve" problem. This is a huge problem for the E.M.

- The assumed magnetic poles are relatively stable at 11^0-17^0 from geographic poles? This is also an unsolved problem. Although it is the total of 7 different vectors as mentioned in the previous conclusions, but $\Delta 6$ and $\Delta 7$ are ascribed as the major players. If the Earth is supposed to be constituted by several major magnet bars, such as continents of America, Euro-Asia, Africa and Antarctica; then the "image" of $\Delta 7$ looks better.

-The figure 19/VI is to guess that the "bundle of field lines" is getting through the Earth core, while the real map in the figures 20 and 21/VI do not confirm so. I guess that the Earth is not a magnet and that the Earth core is inside a type of Faraday's cage where no field line can be connected to. Therefore, the Magnetic Induction Equation (P25 & P197) is not applied to everywhere. Moreover, any attempted calculation for magnetic intensity in the Earth core is needless.

- From Public Health Protection, it points out that the ratio of "+ion condensation/–ion condensation" in city and industrial area often reaches 2000/1000 in every CC of normal air; while it is 1000/1000 or less positive at sea side. This indication also confirms that the land surface is almost positive like what we do in chapter II of part I, but the problem must not be such simple. Thus, the problem of sea/land charge is mentioned before and to be confirmed again that it is a huge question and not solved in this book. It is still unanswered, the contribution of the normal ion ratio to Earth Magnetism.

THIS BOOK CERTAINLY DOES NOT PRESENT EVERYTHING ABOUT EARTH MAGNETISM, AND THE AUTHOR BELIEVES THAT THERE ARE DEFINITELY STILL LOT OF PROBLEMS FOR EARTH MAGNETISM TO BE SOLVED.

Ha Noi, September 09th 2017

Supplementary Images

The following is a map which presents the direction of field lines and the equal-deviation lines as well.

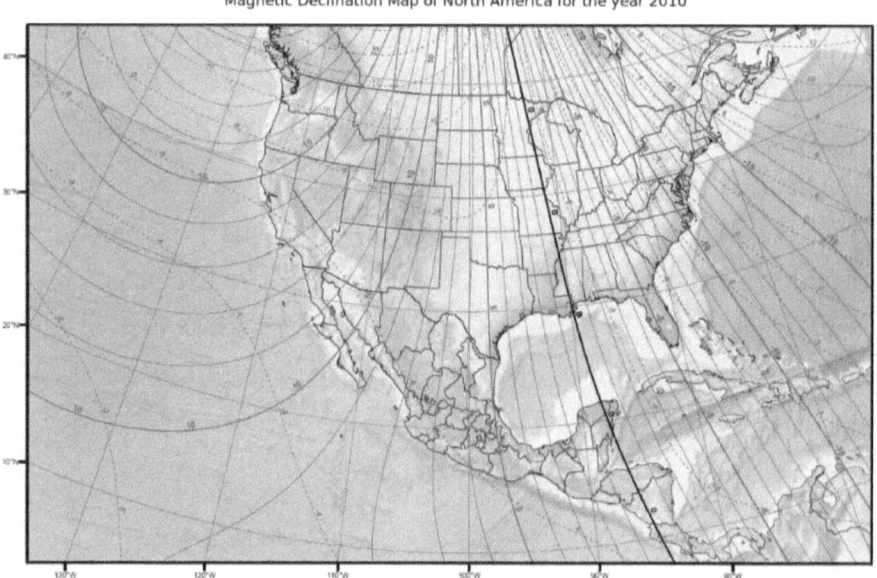

The followings are records of magnetic amplitude on X-wise dated 12th July 2017.

Every record is indicating hourly variableness of magnetic amplitude.

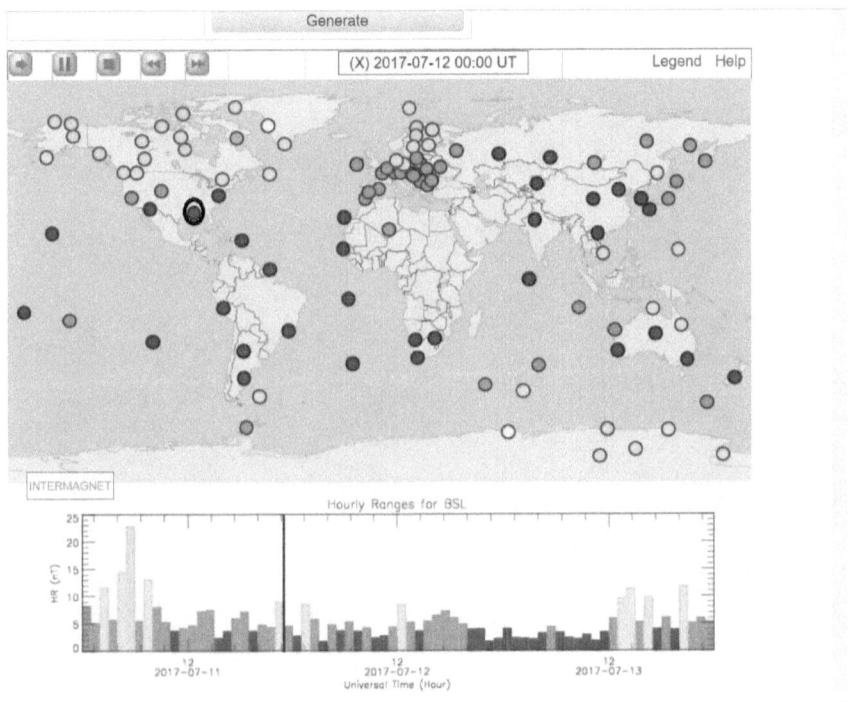

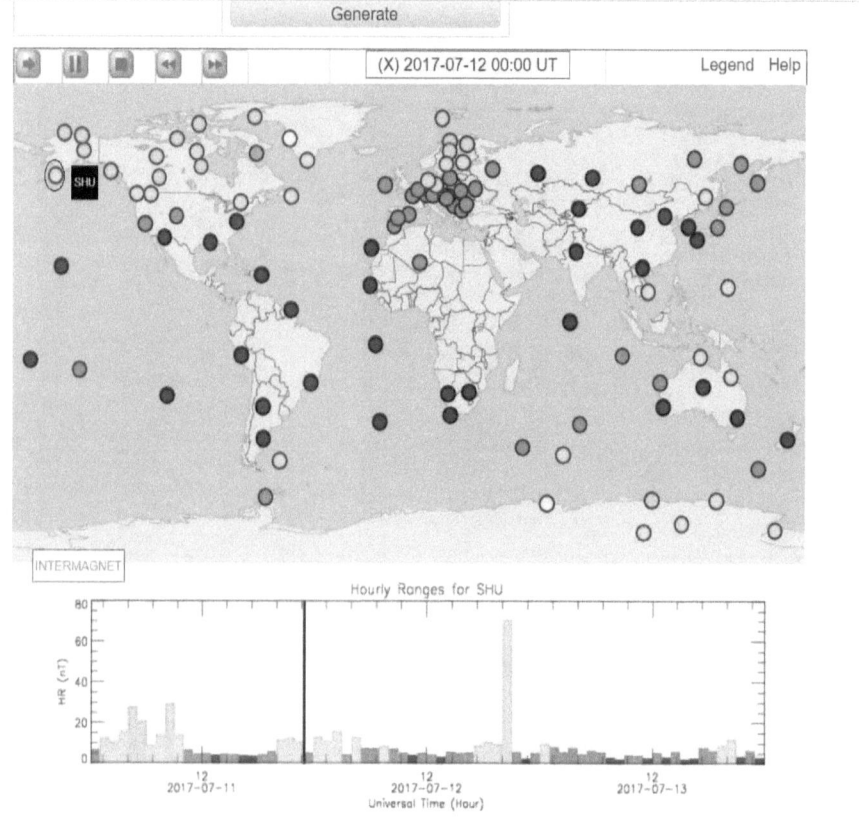

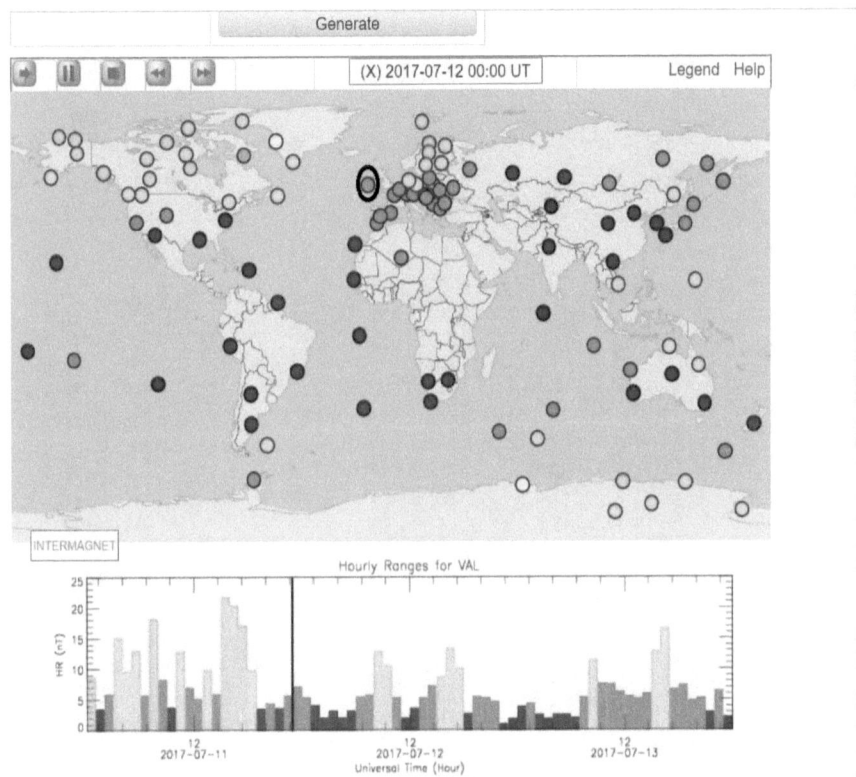

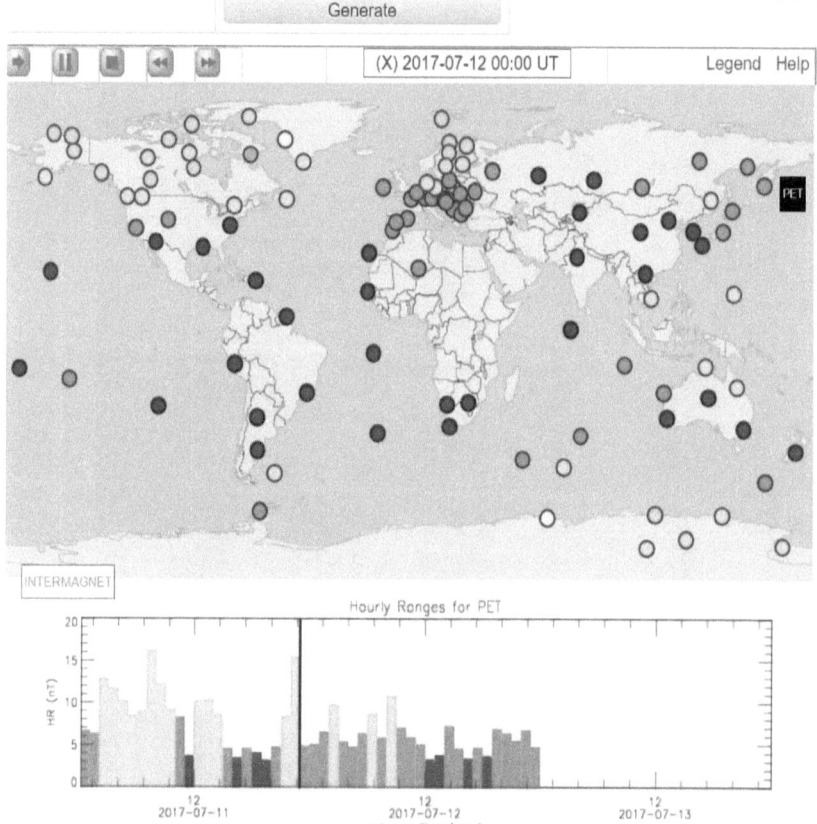

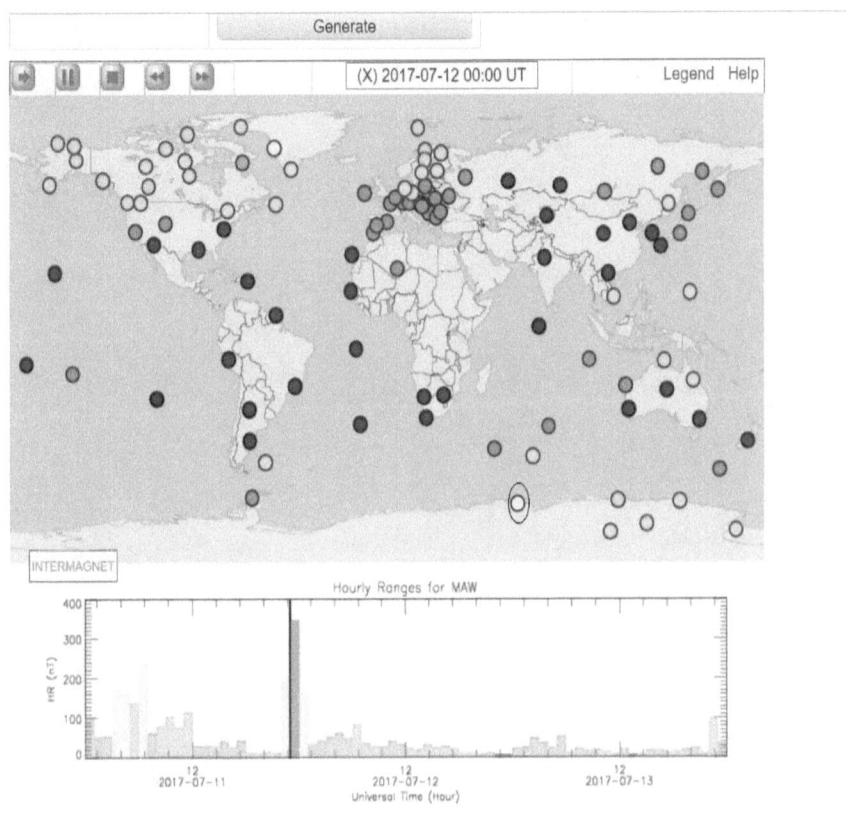

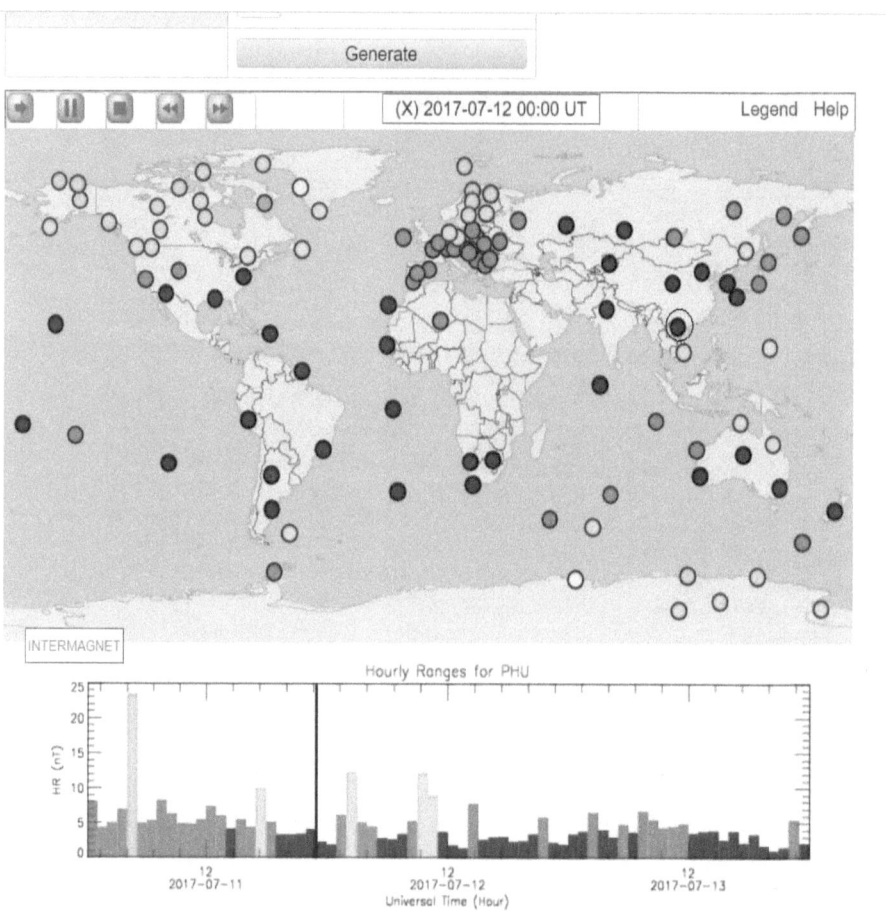

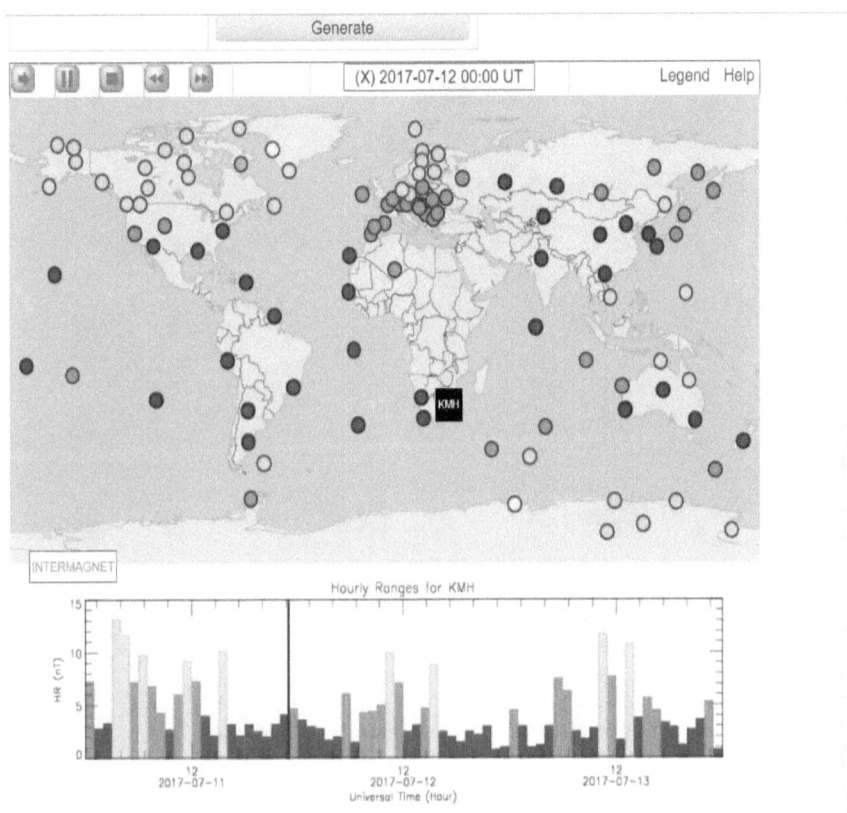

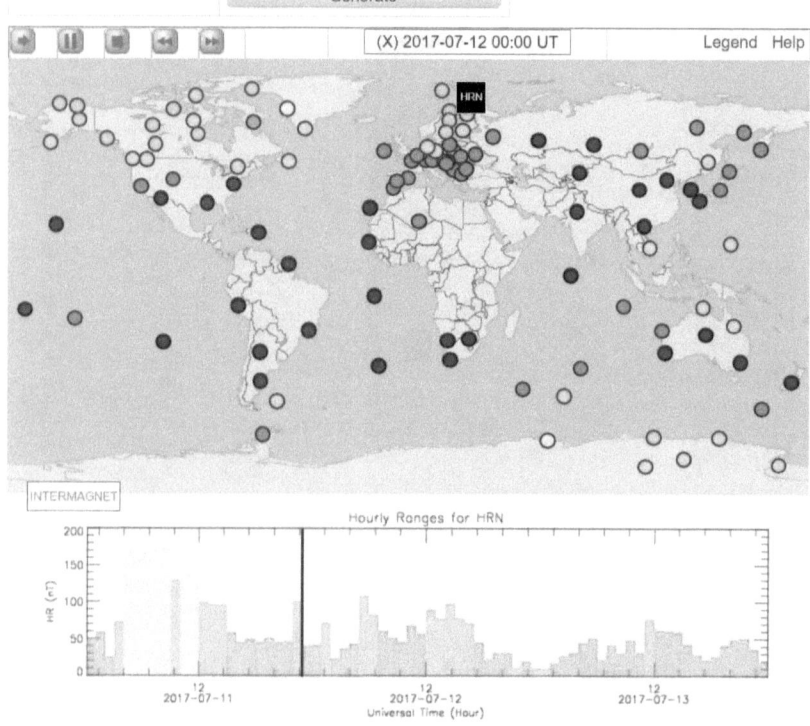

Geomagnetic Information Nodes (GINs)

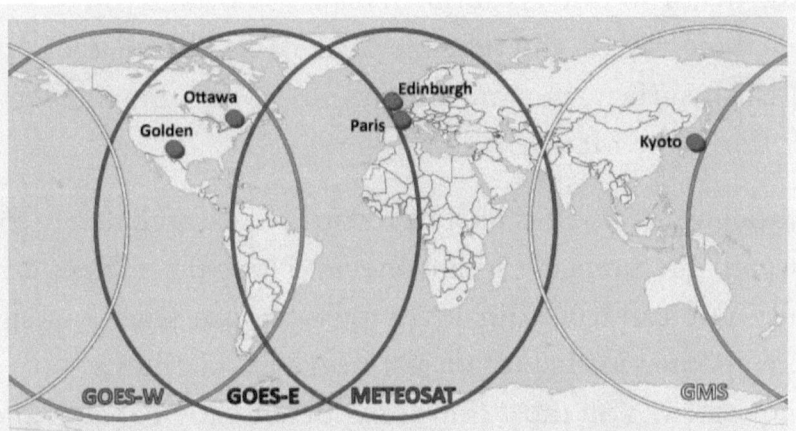

The Geomagnetic Information Nodes (GINs) are the collection points for real-time data within INTERMAGNET. They are connected to the INTERMAGNET observatories by satellite, computer and telephone networks. Minute mean observations of the earth's magnetic field are relayed to the GINs within 72 hours of recording. This time is substantially reduced when observatories are using satellite communications.

The move of electric charge is to make magnetic field, that principle is predominant throughout this book. For the people who love the Earth and pay attention to Earth Magnetism so far, the electric current inside the earth core is the major to discuss about. This book brings you something closer, and tangible to all of us, the charges on the earth surface, in atmosphere and on the only satellite to our Earth-the Moon.

Although this book is likely a record of research, likely for research's reference. But anyone who love the subject can read. You can roll from cover to cover, can scan your sight from chapter to chapter or just read each chapter's conclusions. You must find for yourself a joy to think and an emotion to discuss about Earth Magnetism.

To dedicate this book to Kindle readers, I believe that, like any other researchers, I can bring you several different topics to discuss about the subject: Earth Magnetism.

CONCLUSION

Thank you again for downloading this book on *"EARTH MAGNETISM: Problems with Electric Charges, On Earth, In Atmosphere, In Van Allen Belt and On The Moon"* and reading all the way to the end. I'm extremely grateful.

If you know of anyone else who may benefit from the information presented in this book, please help me inform them of this book. I would greatly appreciate it.

Finally, if you enjoyed this book and feel that it has added value to your study or career in any way, please take a couple of minutes to share your thoughts and post a REVIEW on Amazon. Your feedback will help me to continue to write the kind of Kindle books that helps you get results. Furthermore, if you write a simple REVIEW with positive words for this book on Amazon, you can help hundreds or perhaps thousands of other readers who may want to enhance their knowledge about EARTH MAGNETISM have a chance getting what they need.

Thanks again for your support and good luck!

-- NGUYEN VAN CUONG --

www.ingramcontent.com/pod-product-compliance
Lightning Source LLC
Chambersburg PA
CBHW031619210526
45464CB00004B/1654